# Physical Chemistry

**The INSTANT NOTES series**

Series editor
**B.D. Hames**
*School of Biochemistry and Molecular Biology, University of Leeds, Leeds, UK*

Animal Biology
Ecology
Microbiology
Genetics
Chemistry for Biologists
Immunology
Biochemistry 2nd edition
Molecular Biology 2nd edition
Neuroscience

*Forthcoming titles*
Psychology
Developmental Biology
Plant Biology

**The INSTANT NOTES Chemistry series**
*Consulting editor: Howard Stanbury*

Organic Chemistry
Inorganic Chemistry
Physical Chemistry

*Forthcoming title*
Analytical Chemistry

# Physical Chemistry

## A.G. Whittaker, A.R. Mount
## & M.R. Heal

*Department of Chemistry,
University of Edinburgh, Edinburgh, UK*

A.G. Whittaker, A.R. Mount, and M.R. Heal
*Department of Chemistry, University of Edinburgh, UK*

**Published in the United States of America, its dependent territories and Canada by arrangement with BIOS Scientific Publishers Ltd, 9 Newtec Place, Magdalen Road, Oxford OX4 1RE, UK**

© **BIOS Scientific Publishers Limited 2000**

First published 2000

A CIP catalogue record for this book is available from the British Library.

ISBN 0-387-91619-9   Springer-Verlag New York Berlin Heidelberg   SPIN 10761218

Consulting Editor: Howard Stanbury

Springer-Verlag New York Inc.
175 Fifth Avenue, New York
NY 10010-7858, USA

Production Editor: Paul Barlass
Typeset and illustrated by Phoenix Photosetting, Chatham, Kent
Printed by Biddles Ltd, Guildford, UK, www.biddles.co.uk

# CONTENTS

# ABBREVIATIONS

| | | | |
|---|---|---|---|
| amu | atomic mass unit | LH | left hand |
| Bq | Becquerel | LHS | left hand side |
| Ci | Curie | NMR | nuclear magnetic resonance |
| D | debye | ppm | parts per million |
| emf | electromotive force | RH | right hand |
| ES | enzyme–substrate complex | RHS | right hand side |
| ESR | electron spin resonance | rms | root mean square |
| g | gerade | TMS | tetramethylsilane |
| Gy | gray | u | ungerade |
| LCAO | linear combination of atomic orbitals | | |

# PREFACE

Physical chemistry is an unexpected shock to many university students. From the semi-empirical approaches of the school laboratory, first year undergraduates suddenly find themselves propelled into an unexpected quagmire of definitions and equations. Worse still, although the applicability of the subject is sometimes obvious, studying the behavior of a particle in an infinitely deep well can seem nothing short of farcical on first approach.

In any scientific discipline, a fundamental understanding is more important than learning lists, but this is probably more true in physical chemistry than in other branches of chemistry. Let's be clear from the outset – understanding *is* the key to physical chemistry, but the maelstrom of mathematics often clouds the student's ability to create a comprehensible mental model of the subject.

As the authors of this text, we therefore found ourselves in a paradoxical situation – writing a book containing lists of facts on a subject which isn't primarily about lists of facts. So although this book is primarily a revision text we did not wish it to be merely an encyclopedia of equations and definitions. In order that the conceptual content of the book is given sufficient weight to aid understanding, we have limited the extent of the mathematical treatments to the minimum required of a student. The rigorous arguments which underpin much of physical chemistry are left for other authors to tackle, with our own recommendations for further reading included in the bibliography.

Since our primary aim has been to produce a quick reference and revision text for all first and second year degree students whose studies include physical chemistry, we have recognized that different aspects of the subject are useful in different fields of study. As NMR spectroscopy is to a biochemist's protein study, so is band theory to the solid state chemist, and thermodynamics to the chemical engineer. With this in mind, we have drawn not just on our own teaching experiences, but have consulted with colleagues in the life sciences and in other physical sciences. The rigor of the central themes has not been diluted, but the content hopefully reflects the range of scientists for whom physical chemistry is an important supplement to their main interests.

In organizing the layout of the book, we have aimed to introduce the various aspects of physical chemistry in an order that gives the opportunity for continuous reading from front to back with the minimum of cross-referencing. Thus we start with the basic properties of matter which allows us then to discuss thermodynamics. Thermodynamics leads naturally into equilibria, solutions and then kinetics. The final sections on bonding and spectroscopy likewise follow on from the foundations laid down in the section on quantum mechanics. The background to a range of important techniques is included in the appropriate sections, and once again this reflects the wide application of the subject matter as with, for example, electrophoresis and electro-osmosis.

Whatever your background in coming to this book, our objective has been to use our own perspectives of physical chemistry to aid *your* insight of the subject. Physical chemistry is not the monster that it seems at first, if for no other reason than because a little understanding goes a long way.

We hope that this text contributes to helping you reach the level of understanding you need. Understanding the world around you really is one of the thrills of science.

Finally, we thank Kate, Sue and Janet for all their patience during the preparation of this book.

*M.R. Heal, A.R. Mount, A.G. Whittaker*

# A1 PERFECT GASES

## Key Notes

| | |
|---|---|
| **Gases** | A gas is a fluid which has no intrinsic shape, and which expands indefinitely to fill any container in which it is held. |
| **The perfect gas equations** | The physical properties of a perfect gas are completely described by the amount of substance of which it is comprised, its temperature, its pressure and the volume which it occupies. These four parameters are not independent, and the relations between them are expressed in the gas laws. The three historical gas laws – Boyle's law, Charles' law and Avogadro's principle – are specific cases of the perfect gas equation of state, which is usually quoted in the form $pV = nRT$, where $R$ is the gas constant. |
| **Partial pressure** | The pressure exerted by each component in a gaseous mixture is known as the partial pressure, and is the pressure which that component would exert were it alone in that volume. For a perfect gas, the partial pressure, $p_x$, for $n_x$ moles of each component x is given by $p_x = n_x R T / V$. Dalton's law states that 'the total pressure exerted by a mixture of ideal gases in a volume is equal to the arithmetic sum of the partial pressures'. The quantity $n_A / n_{total}$ is known as the mole fraction of component A, and denoted $x_A$. It directly relates the partial pressure, $p_A$, of a component A, to the total pressure through the expression $p_A = x_A P_{total}$. |

| | | |
|---|---|---|
| **Related topics** | Molecular behavior in perfect gases (A2) | Non-ideal gases (A3) |

**Gases**

A gas is a fluid which has no resistance to change of shape, and will expand indefinitely to fill any container in which it is held. The molecules or atoms which make up a gas interact only weakly with one another. They move rapidly, and collide randomly and chaotically with one another.

The physical properties of an ideal gas are completely described by four parameters which, with their respective SI units are:

- the amount of substance of which it is comprised, $n$, in moles;
- the temperature of the gas, $T$, in Kelvin;
- the pressure of the gas, $p$, in Pascal;
- the volume occupied by the gas, $V$, in m³.

The four parameters are not independent, and the relations between them are expressed in the **gas laws**. The gas laws are unified into a single **equation of state** for a gas which fully expresses the relationships between all four properties. These relationships, however, are based on approximations to experimental observations, and only apply to a **perfect gas**. In what might be deemed a circular argument, a perfect gas is defined as one which obeys the perfect gas equation of state. In practical terms, however, adherence to the perfect gas equation of state requires that the particles which make up the gas are infinitesimally small, and that they interact only as if they were hard spheres, and so perfect gases do not exist. Fortunately, it is found that the behavior of most gases approximates to that of a perfect gas at sufficiently low pressure, with the lighter noble gases (He, Ne) showing the most ideal behavior. The greatest deviations are observed where strong intermolecular interactions exist, such as water and ammonia. The behavior of non-ideal gases is explored in topic A3.

**The perfect gas equations**

Historically, several separate gas laws were independently developed:

**Boyle's law**; $p.V =$ constant    at constant temperature;
**Charles' law**; $V \propto T$    at constant pressure;
**Avogadro's principle**; $V \propto n$    at constant pressure and temperature.

These three laws are combined in the **perfect gas equation of state** (also known as the **ideal gas law** or the **perfect gas equation**) which is usually quoted in the form

$$pV = nRT$$

As written, both sides of the ideal gas equation have the dimensions of energy where R is the **gas constant**, with a value of 8.3145 J K$^{-1}$ mol$^{-1}$. The perfect gas equation may also be expressed in the form $pV_m = RT$, where $V_m$ is the **molar gas volume**, that is, the volume occupied by one mole of gas at the temperature and pressure of interest. The gas laws are illustrated graphically in *Fig. 1*, with lines representing Boyle's and Charles' laws indicated on the perfect gas equation surface.

The gas constant appears frequently in chemistry, as it is often possible to substitute for temperature, pressure or volume in an expression using the perfect gas equation – and hence the gas constant – when developing mathematical expressions.

**Partial pressure**

When two or more gases are mixed, it is often important to know the relationship between the quantity of each gas, the pressure of each gas, and the overall pressure of the mixture. If the ideal gas mixture occupies a volume, $V$, then the pressure exerted by each component equals the pressure which that component would exert if it were alone in that volume. This pressure is called the **partial pressure**, and is denoted as $p_A$ for component A, $p_B$ for component B, etc. With this definition, it follows from the perfect gas equation that the partial pressure for each component is given by:

$$p_x = n_x R T / V$$

where $p_x$ is the partial pressure of $n_x$ moles of component x.

The total pressure exerted by a mixture of ideal gases is related to the partial pressures through **Dalton's law**, which may be stated as,

*'the total pressure exerted by a mixture of ideal gases in a volume is equal to the arithmetic sum of the partial pressures'*.

If a gas mixture is comprised of, for example, $n_A$, $n_B$, and $n_C$ moles of three ideal gases, A, B, and C, then the total pressure is given by:

$$P_{total} = p_A + p_B + p_C = n_A R T / V + n_B R T / V + n_C R T / V = (n_A + n_B + n_C) R T / V$$
$$= n_{total} R T / V$$

where $n_{total}$ is the total number of moles of gas, making this a simple restatement of the **ideal gas law**.

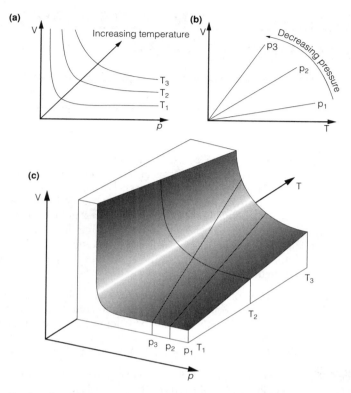

*Fig. 1. Graphical representations of the ideal gas equations. (a) Boyle's law; (b) Charles' law; (c) The surface representing the perfect gas equation. The locations of the lines from (a) and (b) are indicated on the surface.*

The partial pressure of component A divided by the total pressure is given by:

$$p_A/p_{total} = (n_A\,R\,T/V)/(n_{total}\,R\,T/V) = n_A/n_{total}$$

The quantity $n_A / n_{total}$ is known as the **mole fraction** of component A, and is denoted $x_A$ (Topic D1). The advantage of this quantity is that it is easily calculated, and allows ready calculation of partial pressures through the relation:

$$p_A = x_n\,p_{total}$$

# A2 MOLECULAR BEHAVIOR IN PERFECT GASES

## Key Notes

**The kinetic theory of gases**

The kinetic theory of gases is an attempt to describe the macroscopic properties of a gas in terms of molecular behavior. Pressure is regarded as the result of molecular impacts with the walls of the container, and temperature is related to the average translational energy of the molecules. The molecules are considered to be of negligible size, with no attractive forces between them, travelling in straight lines, except during the course of collisions. Molecules undergo perfectly elastic collisions, with the kinetic energy of the molecules being conserved in all collisions, but being transferred between molecules.

**The speed of molecules in gases**

The range of molecular speeds for a gas follows the Maxwell distribution. At low temperatures, the distribution comprises a narrow peak centered at low speed, with the peak broadening and moving to higher speeds as the temperature increases. A useful average, the root mean square (rms) speed, c, is given by $c = (3RT/M)^{1/2}$ where $M$ is the molar mass.

**The molecular origin of pressure**

According to the kinetic theory of gases, the pressure which a gas exerts is attributed to collisions of the gas molecules with the walls of the vessel within which they are contained. The pressure from these collisions is given by $p = (nMc^2)/3V$, where $n$ is the number of moles of gas in a volume $V$. Substitution for c, yields the ideal gas law.

**Effusion**

Effusion is the escape of a gas through an orifice. The rate of escape of the gas will be directly related to the root mean square speed of the molecules. Graham's law of effusion relates the rates of effusion and molecular mass or density of any two gases at constant temperatures:

$$\frac{c_1}{c_2} = \sqrt{\frac{M_2}{M_1}} = \sqrt{\frac{\rho_2}{\rho_1}}$$

**Mean free path**

The mean free path, $\lambda$, is the mean distance travelled by a gas molecule between collisions given by

$$\lambda = \frac{RT}{\sqrt{2}N_A \sigma \rho}$$

where $\sigma$ is the collision cross-section of the gas molecules.

**The collision frequency**

The collision frequency, z, is the mean number of collisions which a molecule undergoes per second, and is given by:

$$z = \frac{\sqrt{2}N_A \sigma \rho c}{RT}$$

**Related topics**    Perfect gases (A1)          Non-ideal gases (A3)

**The kinetic theory of gases**

The gas laws (see Topic A1) were empirically developed from experimental observations. The **kinetic theory of gases** attempts to reach this same result from a model of the molecular nature of gases. A gas is described as a collection of particles in motion, with the macroscopic physical properties of the gas following from this premise. Pressure is regarded as the result of molecular impacts with the walls of the container, and temperature is related to the average translational energy of the molecules.

Three basic assumptions underpin the theory, and these are considered to be true of real systems at low pressure:

1. the size of the molecules which make up the gas is negligible compared to the distance between them;
2. there are no attractive forces between the molecules;
3. the molecules travel in straight lines, except during the course of collisions. Molecules undergo perfectly elastic collisions; i.e. the kinetic energy of the molecules is conserved in all collisions, but may be transferred between them.

**The speed of molecules in gases**

Although the third premise means that the mean molecular energy is constant at constant temperature, the energies, and hence the velocities of the molecules, will be distributed over a wide range. The distribution of molecular speeds follows the **Maxwell distribution of speeds**. Mathematically, the distribution is given by:

$$\frac{dn_s}{N} = f(s)ds = \left(\frac{2}{\pi}\right)^{\frac{1}{2}} \left(\frac{M}{RT}\right)^{\frac{3}{2}} s^2 e^{-ms^2/2RT} \quad ds$$

where $f(s)ds$ is the probability of a molecule having a velocity in the range from $s$ to $s + ds$, $N$ is the number and $M$ is the molar mass of the gaseous molecules. At low temperatures, the distribution is narrow with a peak at low speeds, but as the temperature increases, the peak moves to higher speeds and distribution broadens out (*Fig. 1*).

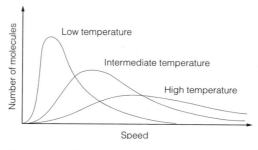

Fig. 1. *The Maxwell distribution of speeds for a gas, illustrating the shift in peak position and distribution broadening as the temperature increases.*

The **most probable speed** of a gas molecule is simply the maximum in the Maxwell distribution curve, and may be obtained by differentiation of the previous expression to give:

$$\text{most probable } s = \left(\frac{2RT}{M}\right)^{\frac{1}{2}}$$

A more useful quantity in the analysis of the properties of gases is the **root mean square (rms) speed**, c. This is the square root of the arithmetic mean of the squares of the molecular speeds given by:

$$c = \left( \frac{3RT}{M} \right)^{\frac{1}{2}}$$

$$\text{where } c = \sqrt{\frac{s_1^2 + s_2^2 + \ldots s_N^2}{N}} = \sqrt{\frac{\sum_{i=1}^{N} s_i^2}{N}}$$

The rms speed is always greater than the most probable speed. For oxygen molecules at standard temperature, the most probable speed is 393 m s$^{-1}$ and the root mean square speed is 482 m s$^{-1}$.

**The molecular origin of pressure**

In the kinetic theory of gases, the pressure which a gas exerts is attributed to collisions of the gas molecules with the walls of the vessel within which they are contained. A molecule colliding with the wall of the vessel will change its direction of travel, with a corresponding change in its momentum (the product of the mass and velocity of the particle). The force from the walls is equal to the rate of change of momentum, and so the faster and heavier and more dense the gas molecules, the greater the force will be. The equation resulting from mathematical treatment of this model may be written as:

$$p = \frac{nMc^2}{3V}$$

where $n$ is the number of moles of gas in a volume $V$ (i.e. the density). This equation may be rearranged to a similar form to that of the ideal gas law: $PV = n\, Mc^2/3$.

Substituting for c, yields $PV = n\, m(3RT/m)/3 = nRT$, i.e. the ideal gas law.

Alternatively, we may recognize that the value $\frac{1}{2}Mc^2$ represents the rms kinetic energy of the gas, $E_{kinetic}$, and rewrite the equation to obtain the **kinetic equation for gases**:

$$pV = \frac{2nE_{kinetic}}{3}$$

**Effusion**

Effusion is the escape of a gas through an orifice. The rate of escape of the gas is directly related to c:

$$c = \sqrt{\frac{3RT}{M}} = \sqrt{\frac{3pV}{nM}} = \sqrt{\frac{3p}{\rho}}$$

where $\rho$ is the density of the gas. For two gases at the same temperature and pressure, for example nitrogen and hydrogen, it follows that the ratio of the velocities is given by:

$$\frac{c_{H_2}}{c_{N_2}} = \sqrt{\frac{M_{N_2}}{M_{H_2}}} = \sqrt{\frac{\rho_{N_2}}{\rho_{H_2}}}$$

This is **Graham's law of effusion**.

**Mean free path**

Gas particles undergo collisions with other gas particles in addition to colliding with the walls. The mean distance travelled by a gas molecule between these random collisions is referred to as the **mean free path, $\lambda$**. If two molecules are regarded as hard spheres of radii $r_A$ and $r_B$, then they will collide if they come within a distance $d$ of one another where $d = r_A + r_B$. (see Topic F3, *Fig.* 2). The area circumscribed by this radius, given by $\pi d^2$, is the **collision cross-section, $\sigma$**, of the molecule. As molecules are not hard spheres, the collision cross-section will deviate markedly from this idealized picture, but $\sigma$ still represents the effective physical cross-sectional area within which a collision may occur as the molecule travels through the gas. The mean free path decreases with increasing value of $\sigma$, and with increasing pressure, and is given by:

$$\lambda = \frac{RT}{\sqrt{2}N_A\sigma\rho} = \frac{k_B T}{\sqrt{2}\sigma\rho}$$

where **Avogadro's constant, $N_A$**, converts between molar and molecular units ($R = N_A k_B$, where $k_B$ is the **Boltzmann Constant**).

**The collision frequency**

The **collision frequency, $z$**, of molecules in a gas is the mean number of collisions which a molecule will undergo per second. The collision frequency is inversely related to the time between collisions and it therefore follows that $z$ is inversely proportional to the **mean free path** and directly proportional to the speed of the molecule i.e. $z = c / \lambda$,

$$z = \frac{\sqrt{2}N_A\sigma\rho c}{RT}$$

# A3 NON-IDEAL GASES

## Key Notes

**Non-ideal gases**

Real gases at moderate and high pressures do not conform to the ideal gas equation of state, as intermolecular interactions become important. At intermediate pressures, attractive forces dominate the molecular interactions, and the volume of the gas becomes lower than the ideal gas laws would predict. At higher pressures, repulsive forces dominate the intermolecular interactions. At high pressure, the volume of all gases is larger than the ideal gas law predicts, and they are also much less compressible.

**The virial equation**

The virial equation is a mathematical approach to describing the deviation of real gases from ideal behavior by expanding the ideal gas equation using powers of the molar volume, $V_m$:

$$pV_m = RT(1 + \frac{B}{V_m} + \frac{C}{V_m^2} + \frac{D}{V_m^3} + ...)$$

The virial coefficients B, C, D, etc. are specific to a particular gas, but have no simple physical significance. A gas at low pressure has a large molar volume, making the second and subsequent terms very small, and reduces the equation to that of the perfect gas equation of state.

**The van der Waals equation of state**

This is a modification of the perfect gas equation which allows for the attractive and repulsive forces between molecules. The equation has the form $(V - nb)(p + a(n/V)^2) = nRT$. The van der Waals parameters, $a$ and $b$, convey direct information about the molecular behavior. The term $(V - nb)$ models the repulsive potential between the molecules, and the term $(p + a(n/V)^2)$ compensates for the attractive potential. At high temperatures and low pressures, the correction terms become small compared to $V$ and $T$ and the equation reduces to the perfect gas equation of state.

**Related topics**

Perfect gases (A1)

Molecular behavior in perfect gases (A2)

**Non-ideal gases**

Ideal gases are assumed to be comprised of infinitesimally small particles, and to interact only at the point of collision. At low pressure, the molecules in a real gas are small relative to the mean free path, and sufficiently far apart that they may be considered only to interact close to the point of collision, and so comply with this assumption. Because the intermolecular interactions become important for real gases at moderate and high pressures, they are **non-ideal gases** and they no longer conform to the ideal gas laws.

At intermediate pressures, attractive forces dominate the molecular interactions, and the volume of the gas becomes lower than the ideal gas laws would predict. As progressively higher pressures are reached, the molecules increase their proximity to one another and repulsive forces now dominate the inter-molecular interactions. At high pressures, all gases have a higher volume than the ideal gas law predicts, and are much less compressible.

The **compression factor**, $Z$ expresses this behavior, and is commonly plotted as a function of pressure. It is defined as:

$$Z = \frac{pV_m}{RT}$$

where $V_m$ is the **molar volume**.

$Z$ is equal to 1 at all pressures for a gas which obeys the ideal gas law, and it is found that all gases tend to this value at low pressure. For all real gases, $Z$ is greater than 1 at high pressure, and for many gases it is less than 1 at inter-mediate pressures. The plot of $Z$ as a function of $p$ is shown in *Fig. 1*. Note that an equivalent plot of the product $pV$ as a function of pressure at constant temperature is commonly used, and takes an almost identical form.

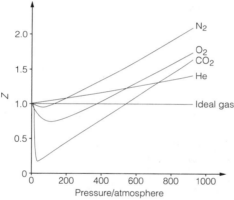

*Fig. 1. Deviation of Z from ideality as a function of pressure.*

Numerous attempts have been made to modify the perfect gas equation of state in order to describe real gases. The two most significant equations are the virial equation and the Van der Waals equation of state.

**The virial equation**

The **virial equation** is primarily a mathematical attempt to describe the devia-tion from ideality in terms of powers of the molar volume, $V_m$. It takes the form:

$$pV_m = RT(1 + \frac{B}{V_m} + \frac{C}{V_m^{\;2}} + \frac{D}{V_m^{\;3}} + ...)$$

The coefficients $B$, $C$, $D$, etc. are the **virial coefficients**. The expression converges very rapidly, so that $B > C > D$, and the expression is usually only quoted with values for $B$ and $C$ at best. For a gas at low pressure, $V_m$ is large, making the second and subsequent terms very small, reducing the equation to that of the perfect gas equation of state. For an ideal gas, $B$, $C$, $D$, etc. are equal to zero, and the equation again reduces to that for an ideal gas. Although the virial equation provides an accurate description of the behavior of a real gas, the fit is empirical, and the coefficients $B$, $C$, $D$, etc., are not readily related to the molecular behavior.

**The van der Waals equation of state**

The **van der Waals equation of state** attempts to describe the behavior of a non-ideal gas by accounting for both the attractive and repulsive forces between molecules. The equation has the form:

$$\left( p + a\left(\frac{n}{V}\right)^2 \right)(V - bn) = nRT$$

The coefficients $a$ and $b$ are the **van der Waals parameters**, and have values which convey direct information about the molecular behavior.

The term $(V - nb)$ models the repulsive potential between the molecules. This potential has the effect of limiting the proximity of molecules, and so reducing the available volume. The excluded volume is proportional (through the coefficient, $b$) to the number of moles of gas, $n$. The term $(p + a(n/V)^2)$ reflects the fact that the attractive potential reduces the pressure. The reduction in pressure is proportional to both the strength and number of molecular collisions with the wall. Because of the attractive potential, both of these quantities are reduced in proportion to the density of the particles $(n/V)$, and the overall pressure reduction is therefore $a(n/V)^2$ where $a$ is a constant of proportionality.

At high temperatures and large molar volumes (and therefore low pressures), the correction terms become relatively unimportant compared to $V$ and $T$ and the equation reduces to the perfect gas equation of state.

The van der Waals equation shows one feature which does not have physical significance. If the equation is plotted as a function of pressure against volume, then all $p$-$V$ isotherms below a **critical temperature** display loops, within which the volume of the gas apparently decreases with increasing pressure. These **van der Waals loops** are part of a two-phase region within which gas and liquid coexist, a proportion of the gas having condensed to a liquid phase. To represent physical reality, a horizontal line replaces the loops (*Fig. 2*). Along this line,

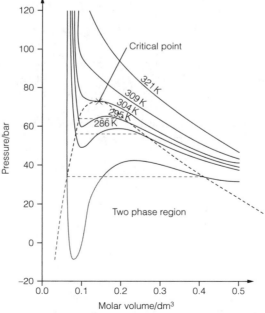

*Fig. 2.   Plots of the van der Waals equation of state, showing the van der Waals loops replaced by horizontal lines, and the critical point.*

varying proportions of gas and liquid coexist, with the right-hand end representing pure gas, and the left-hand end pure liquid.

The **critical pressure,** $P_c$ and **critical temperature,** $T_c$ are the pressure and temperature above which this two phase region no longer exists. The critical temperature may be calculated from the Van der Waals equation by differentiation, recognizing that at the **critical point** on *Fig. 2*, there is a horizontal point of inflexion in the pressure–volume curve. This method gives

$$T_c = \frac{8a}{27bR} \quad \text{and} \quad P_c = \frac{a}{27b^2}$$

# A4 Liquids

## Key Notes

**Structure of liquids**

Liquids have a limited degree of short-range order, but virtually no long-range order, and are most adequately described in terms of a radial distribution function – the probability of finding a neighbor at a given radial distance. The radial distribution function displays a temperature dependence which correlates with the effects of temperature on the structure. Generally, increasing temperature increases the radial distance of the peaks in the radial distribution function, corresponding to the thermal expansion of the liquid. The peak intensities also become reduced, as increasing temperature leads to a more chaotic and dynamic liquid structure.

**Viscosity**

Viscosity characterizes the motion of fluids in the presence of a mechanical shear force. A fluid passing through a capilliary experiences a retarding force from the walls of the tube, resulting in a higher velocity along the central axis than at the walls. For any given small bore capilliary, it is found that a specified volume of fluid, of density, $\rho$, flows through the capilliary in a time, $t$, given by the

relation: $\dfrac{\eta}{\rho} \propto t$.

It is convenient to define a quantity known as the frictional coefficient, $f$, which is directly related to molecular shapes through Stoke's law. In the ideal case of spherical particles this may be expressed simply as $f = 6\pi\eta r$.

**Diffusion**

The tendency of a solute to spread evenly throughout the solvent in a series of small, random jumps is known as diffusion. The fundamental law of diffusion is Fick's first law. In the ideal case of diffusion in one dimension, the rate of diffusion of $dn$ moles of solute, $dn/dt$, across a plane of area $A$, is proportional to the diffusion coefficient, $D$, and the *negative* of the concentration gradient, $-dc/dx$:

$$dn/dt = -DA\, dc/dx$$

The diffusion coefficient for a spherical molecule, of radius $r$, is related to the viscosity of the solvent through $D = kT/(6\pi\eta r)$. If it is assumed that the molecule makes random steps, then D also allows calculation of the mean square distance, $x^2$, over which a molecule diffuses in a time, $t$, by the relation $\overline{x^2} = 2Dt$.

**Surface tension**

In the absence of other forces, the free energy of a liquid is minimized when it adopts the minimum surface area possible. The free energy change in a surface of area $A$, depends on the surface tension of the liquid, $\gamma$, and is given by $dG = \gamma\, dA$. For a gas cavity of radius $r$ within a liquid, the pressure difference between the inside and the outside of the cavity is given by $\Delta p = p_{gas} - p_{liquid} = 2\gamma/r$.

For bubbles, the presence of two surfaces doubles the pressure differential between the inside and outside of the bubble for a given radius: $\Delta p = p_{inside} - p_{outside} = 4\gamma/r$.

**Surfactants**

Surfactants are chemical species with a tendency to accumulate at surfaces, and tend to lower the surface tension of a liquid. Most surfactants are composed of a hydrophilic head and a hydrophobic tail. Assuming the solvent to be water or another polar solvent, the conflicting requirements of the two groups are met at the surface, with the head remaining in the solvent and the tail pointing out of the solvent. Above a critical concentration and above the Krafft temperature, surfactant molecules may not only accumulate at the surface, but may also form micelles. Micelles are clusters of between some tens and some thousands of surfactant molecules whose tails cluster within the micelle so as to maximize interactions between the tails, leaving a surface of solvated hydrophilic heads.

**Liquid crystals**

Materials in a superficially liquid state which retain most of their short-range order, and some of their long-range order are no longer solid nor are they truly liquid, and are termed liquid crystals. Liquid crystals tend to be formed from molecules which are highly anisotropic, with rod, disk, or other similar shapes. In the smectic phase, molecules are aligned parallel to one another in regular layers. In the nematic phase, the molecules are aligned parallel to one another, but are no longer arranged in layers, and in the cholosteric phase ordered layers of molecules are aligned with respect to one another within each layer, but the layers are no longer ordered with respect to one another.

**Related topics**

Molecular behavior in perfect gases (A2)

Molecular aspects of ionic motion (E7)

**Structure of liquids**

The structure of a liquid is intermediate between that of a solid (see Topic A5) and a gas (see Topic A3). The molecules in a liquid have sufficient energy to allow relative motion of its constituent molecules, but insufficient to enable the truly random motion of a gas. Liquids have a limited degree of short-range order, but virtually no long-range order, and in contrast to a solid, a liquid cannot be adequately described in terms of atomic positions. They are better described in terms of a radial distribution function, since there is only the probability of finding a neighbor at a given radial distance, rather than the certainty of a neighbor at a fixed point.

The radial distribution function shows that some degree of short-range order exists in liquids, insofar as there are typically three or four distinct radii at which there is relatively high probability of finding a neighbor. This variability rapidly diminishes, and at large distances, the probability is approximately uniform in all directions (**isotropic**). The radial distribution function for an idealized liquid is shown in *Fig. 1*.

The temperature dependence of the radial distribution function reflects the effects of temperature on the structure. Generally, increasing temperature increases the radial distance of the peaks in the radial distribution function, corresponding to the thermal expansion of the liquid. The peak intensities also

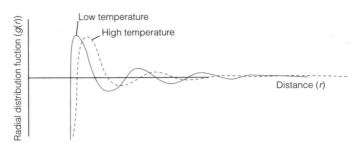

Fig. 1.   *Radial distribution function for an idealized liquid.*

become reduced, as increasing temperature leads to a more chaotic and dynamic liquid structure.

**Viscosity**

Viscosity characterizes the motion of fluids in the presence of a mechanical shear force. The simplest approach takes two slabs of fluid, each of area $A$, a distance $d$ apart, within a larger sample of the fluid, of viscosity $\eta$. A shear force $F$ is applied to one slab, so as to cause the slabs to move at a relative velocity $v$. The force is then given by:

$$F = \frac{Av}{d}\eta$$

A fluid passing through a capillary experiences a retarding force from the walls of the tube, resulting in a higher velocity along the central axis than at the walls. For any given capillary the time, $t$, taken by a specified volume of fluid of density, $\rho$, to pass through the capillary is related to the viscosity through the relation:

$$\eta / \rho \propto t.$$

Rather than measuring absolute viscosity, it is more convenient to measure the time taken for a specific volume of a liquid standard to pass through a capillary, and to compare this with the time required for the same volume of the fluid of interest, whence:

$$\frac{\eta_{sample}}{\rho_{sample} t_{sample}} = \frac{\eta_{standard}}{\rho_{standard} t_{standard}}$$

It is convenient to define a quantity known as the **frictional coefficient**, $f$, which is given simply by $f = A\eta/d$. This quantity can be measured relatively easily, but is directly related to molecular shapes through **Stoke's law**:

$$f = 6\pi\eta r \{F(a,b,c)\}$$

$r$ is the **effective radius** of the molecule, and represents the radius of a sphere with the same volume as that of the molecule. $F(a,b,c)$ is a complex shape-dependent function of the molecule's dimensions. Fitting of this expression to experimental data allows determination of molecular shapes. For spherical molecules, $F(a,b,c) = 1$.

**Diffusion**

When a solute is present in a solvent, then the tendency of that solute is to spread evenly throughout the solvent in a series of small, random jumps. This thermally energized process is known as **diffusion**. The fundamental law of diffusion is **Fick's first law**. The rate of diffusion of $dn$ moles of solute, $dn/dt$, across a plane of area $A$, is proportional to the **diffusion coefficient**, $D$, and the *negative* of the concentration gradient, $-dc/dx$, thus:

$$\frac{dn}{dt} = -D\,A\,\frac{dc}{dx}$$

The diffusion coefficient for a spherical molecule, of radius $r$, is very simply related to the viscosity of the solvent:

$$D = k_B T / (6\,\pi\,\eta\,r)$$

where $k_B$ is the **Boltzmann constant** and $T$ is the temperature. Alternatively, if it is assumed that the molecules take steps of length $\lambda$ in time $\tau$, $D$ is also given by $D = \lambda^2/2\tau$. If it is further assumed that the molecule makes random steps, then $D$ also allows calculation of the mean square distance, $\overline{x^2}$, over which a molecule diffuses in a time, $t$:

$$\overline{x^2} = 2Dt$$

**Surface tension**

The effect of **intermolecular forces** in a liquid results in the **free energy** of a liquid being minimized when the maximum number of molecules are completely surrounded by other molecules from the liquid. More familiarly, this implies that a liquid will tend to adopt the minimum surface area possible. Surfaces are higher energy states than the bulk liquid and $G_{surface}$, the free energy of a surface of area $A$, is defined by

$$G_{surface} = \gamma\,A$$

where $\gamma$ is the **surface tension** of the liquid. Typically, $\gamma$ ranges between $47 \times 10^{-2}$ N m$^{-1}$ for mercury down to $1.8 \times 10^{-2}$ N m$^{-1}$ for a liquid with relatively small intermolecular interactions such as pentane. It follows that small changes of surface area $dA$ result in an amount of work $dG$ given by $dG = \gamma\,dA$.

In a gas **cavity** (a volume which is wholly contained by the liquid), the effect of the surface tension is to minimize the liquid surface area, and hence the volume of the cavity. The outward pressure of the gas opposes this minimization. For a cavity of radius $r$, the pressure difference between the inside and the outside of the cavity, $\Delta p$, is given by:

$$\Delta p = p_{gas} - p_{liquid} = 2\gamma/r$$

The inverse relationship in $r$ means that gas within a small cavity must be at a higher pressure than the gas in a large cavity for any given liquid. At very small radii, the pressure difference becomes impractically large, and is the reason why cavities cannot form in liquids without the presence of **nucleation sites**, small cavities of gas in the surface of particles within the liquid.

For **bubbles** – a volume of gas contained in a thin skin of liquid – two surfaces now exist, and the pressure differential between the inside and outside of the bubble becomes:

$$\Delta p = p_{inside} - p_{outside} = 4\gamma/r$$

**Surfactants**

**Surfactants** are chemical species which have a tendency to accumulate at surfaces, and tend to lower the surface tension of the liquid. A familiar example is a cleaning detergent, which is comprised of a **hydrophilic head**, such as -SO$_3^-$

or -CO$_2^-$, and a **hydrophobic tail**, which is usually comprised of a long-chain hydrocarbon. In water or other polar solvents, the head group has a tendency to solvation, whilst the tail adopts its lowest free energy state outside the solvent. The compromise between these conflicting requirements is met at the surface, with the head remaining in the solvent and the tail pointing out of the solvent. Above a **critical surfactant concentration** and above the **Krafft temperature**, surfactant molecules may not only accumulate at the surface, but may also form micelles. **Micelles** are clusters of between some tens and some thousands of surfactant molecules within a solvent whose tails cluster within the micelle so as to maximize the interactions of the tails, leaving a surface of solvated hydrophilic heads. Where surfactants are added to water in the presence of greases or fats, the tails may solvate in the fatty material, leaving a surface of hydrophilic heads. The effect is to dissolve the grease in the water, and is the reason for the cleaning properties of detergents and soaps.

**Liquid crystals**

A material may melt from the **crystalline state** (see Topic A5) into a superficially liquid state, and yet retain most of the short-range order, and some of the long-range order so that it cannot be considered to be a true liquid. Such materials are neither wholly solid nor wholly liquid, and are termed **liquid crystals**. Liquid crystals tend to be formed from molecules which are highly anisotropic, with rod, disk, or other similar shapes. Several possible phases are adopted by liquid crystals, depending upon the nature and degree of order which is present, and these are illustrated in *Fig. 2*. The most ordered phase is the **smectic phase**, in which molecules are aligned parallel to one another in regular layers. In the **nematic phase**, the molecules are aligned parallel to one another, but are no longer arranged in layers. The **cholosteric phase** is characterized by ordered layers in which the molecules are aligned with respect to one another within each layer, but the layers are no longer ordered with respect to one another. In all these phases, the material flows like a liquid, but exhibits optical properties akin to those of a solid crystal. The typical operating range for liquid crystals is between –5°C and 70°C. Below this range, the material is a true crystalline solid, and above this range, all order is lost and the material behaves as an **isotropic liquid**.

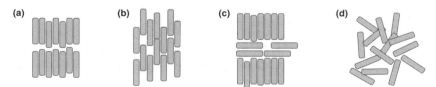

*Fig. 2.   Liquid crystal phases, illustrated with idealized rod-shaped molecules. (a) Smectic phase; (b) Nematic phase; (c) Cholosteric phase; (d) Isotropic liquid.*

# A5 CRYSTALLINE SOLIDS

## Key Notes

**Crystalline solids**

Solids may be broadly grouped into two categories, amorphous and crystalline. Crystalline materials are characterized by highly ordered packing of molecules, atoms or ions. This order allows relatively easy structural studies. Seven crystal systems exist in three-dimensional crystals, from which all possible crystal morphologies may be generated. The deviation from these crystal systems which real crystals exhibit is primarily due to the different growth rates of each crystal face.

**Unit cells**

A crystalline material is composed of an array of identical units. The smallest unit which possesses all of the properties of the crystal is the unit cell. From a unit cell, the entire crystal may be built up by allowing a simple translation operation parallel to any of the three unit cell axes. In principle, there are an almost infinite number of possible unit cells, but it is customary to choose a unit cell which exhibits the symmetry properties of the entire lattice, within the minimum volume, and with angles as close as possible to 90°. In three-dimensional crystals, the 14 Bravais lattices are sufficient to account for all possible unit cells.

**Lattice planes**

In addition to the planes which are parallel to the cell axes, an ordered array also contains an infinite number of sets of parallel planes containing the basic motif. The interplanar distances are of primary importance in diffraction studies, and the Miller indices provide the most useful method for discussing the physical attributes of particular sets of lattice planes. Most notably, Miller indices allow interplanar distances to be readily calculated, which ultimately allows convenient analysis of X-ray and neutron diffraction measurements.

**Related topics**

Diffraction by solids (A6)

**Crystalline solids**

Solids may be loosely categorized into two groups. **Amorphous solids** have no long-range order in their molecular or atomic structure. By their nature they are not easily studied, since the powerful analytical methods which are described in Topic A6, for example, are not applicable to such disordered structures. In contrast **crystalline solids** which consist of ordered three-dimensional arrays of a structural motif, such as an atom, molecule or ion. This internal order is reflected in the familiar macroscopic structure of crystalline materials, which typically have highly regular forms with flat crystal faces. It is this order and regularity which enables much simpler structural studies of crystalline materials.

The huge range and variety of crystal morphologies which are observed might signify that there are a correspondingly wide range of crystal groups into which these shapes may be categorized. In fact, it turns out that by grouping the crystals according to the angles between their faces and the equivalence of the growth along each axis, only seven **crystal systems** are required to encompass all possible crystal structures (*Fig. 1*). A crystal of a material such as sodium chloride, for example, clearly exhibits three equivalent perpendicular axes, and so belongs to the cubic crystal system, whereas crystals of γ-sulfur possess two perpendicular axes with a third axis at an obtuse angle to these, and so belongs to the monoclinic system. The variety of crystal forms which result from this limited number of crystal systems is primarily a result of the different rates at which different crystal faces grow.

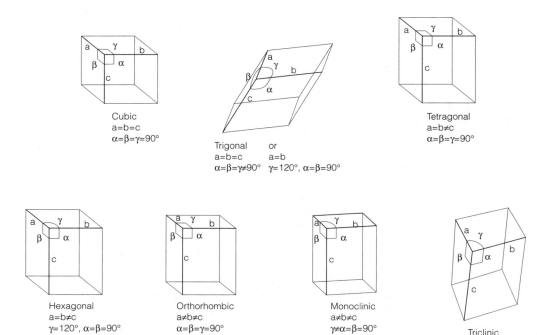

*Fig. 1. The seven crystal systems.*

**Unit cells**

The **structural motif** (i.e. the atom or molecule) which makes up a crystalline solid may adopt any one of a large range of distinct orderly structures. It is precisely *because* crystalline materials are ordered infinite three-dimensional arrays that their study is possible, since the problem may be reduced to the properties of a small portion of the array. Since the crystal contains a repeated structure it is possible to locate a basic unit within the array which contains all the symmetry properties of the whole assembly. This basic building block is referred to as the **unit cell**, and is the smallest unit which contains all the components of the whole assembly. It may be used to construct the entire array by repetition of simple translation operations parallel to any of its axes. By convention, the three unit cell edge lengths are denoted by the letters a, b and c. Where a, b and c are identical, all three edges are denoted a, and where two are identical, these are denoted a, with the third denoted by c. The angles between the axes are likewise denoted α, β and γ.

Because there are an infinite number of possible unit cells for any given array, several principles govern unit cell selection:

(i)   the edges of the unit cell should be chosen so as to be parallel with symmetry axes or perpendicular to symmetry planes, so as to best illustrate the symmetry of the crystal;

(ii)  the unit cells should contain the minimum volume possible. The unit cell lengths should be as short as possible and the angles between the edges should be as close to 90° as possible;

(iii) where angles deviate from 90°, they should be chosen so as to be all greater than 90°, or all smaller than 90°. It is preferable to have all angles greater than 90°;

(iv)  the origin of the unit cell should be a geometrically unique point, with centres of symmetry being given the highest priority.

*Fig. 2* illustrates some of these points for a two-dimensional lattice. Some possible unit cells for the rhombohedral array of points (*Fig. 2a*) are shown in *Fig. 2b*, and whilst repetition of any one of these unit cells will generate the entire lattice, only one of them complies with (i), (ii) and (iii) above. *Fig. 2c* and *Fig. 2d* illustrate principle (iv) above. Taking the most appropriate unit cell for this array its position is selected so as to place its origin at a geometrically unique point (i.e. on a lattice element). Therefore, whilst the unit cell shown in *Fig. 2c* is permissible, the unit cell shown in *Fig. 2d* is preferred for its compliance with principle (iv).

It is worth noting that these principles are only guidelines, and other considerations may occasionally mean that it is beneficial to disregard one or more of them. One might, for example, select a unit cell so as to better illustrate a property of the crystal, or so as to allow easier analysis through diffraction methods (see Topic A6).

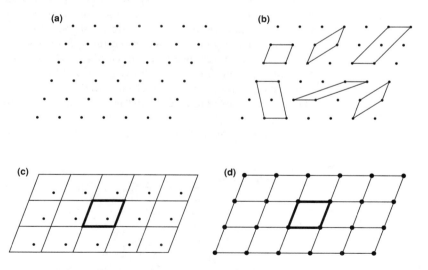

Fig. 2.   *Possible choices of unit cells for a regular two-dimensional array of atoms.*

Unit cells themselves have a limited number of symmetry properties, since they must be capable of packing together to completely fill an area or volume of space. For a two-dimensional array, this restriction means that the possible unit cells must possess 1-, 2-, 3-, 4- or 6-fold rotational symmetry, with no other possible symmetries. Five- or seven-fold unit cell symmetries, for example, would require tiling pentagons or heptagons on a flat surface, an operation which is easily demonstrated to be impossible. Two-dimensional packing is therefore limited to five basic types of unit cell.

For three-dimensional crystalline arrays, the same fundamental arguments apply, and fourteen basic unit cells exist which may be packed together to completely fill a space. These are referred to as the **Bravais lattices**. There are seven **primitive** unit cells (denoted P), with **motifs** placed only at the vertices. Two **base centered** unit cells (C) may be formed from these primitive unit cells by the addition of a motif to the center of two opposing unit cell faces. Two **face centered** unit cells (F) result from adding a motif to all six face centers. Three **body centered** unit cells (B) are generated by placing a motif at the center of the unit cell.

Since each of the body centered and face centered unit cells are generated by adding atoms to the primitive unit cells, it might be supposed that other unit cells may be generated, such as, for example, face-centered tetragonal. In fact, all such attempts to generate new unit cells inevitably generate one of the 14 Bravais lattices.

It should be appreciated that the symmetry of the unit cell is *not* necessarily related to the symmetry of its motifs, only to its packing symmetry. Hence, it is perfectly possible for ferrocene, a molecule with five-fold rotational symmetry, to pack in a structure with a hexagonal unit cell.

**Lattice planes**  X-ray and **neutron diffraction** techniques for structural analysis (see Topic A6) can only be interpreted by understanding how the diffraction patterns result from the internal arrangement of the atoms or molecules. **Lattice planes**, although not a truly rigorous approach are, nevertheless, a very useful aid to understanding diffraction. As with unit cells, matters are simplified by considering a two-dimensional lattice of atoms or molecules, whilst recognizing that the arguments may be extended into three dimensions at a later point. Consider, for example, a two-dimensional lattice (*Fig. 3*). This array clearly contains rows of points parallel to the a and b axes, but in addition to these rows, however, other sets of rows may also be selected. In three dimensions, these rows become planes (*Fig. 4*), but are constructed in a similar fashion to those in two dimensions.

The different lattice rows or planes are formally distinguished by their **Miller**

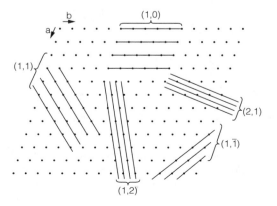

Fig. 3.   *Lattice rows in a two-dimensional lattice.*

indices. Taking the two-dimensional example, it is possible to ascribe a unique set of intersect points for each set of rows. Thus in *Fig. 3*, the rows beginning at the top left-hand edge passes through the next point at 1a and 1b. For other rows, the intersect is at, for example, (3a, 2b). As the distances can always be written in terms of the length of the unit cell, and the notation is simplified by referring to these intersects as (1,1) and (3,2) respectively. The Miller indices are obtained by taking the reciprocal of the intersects. Where the reciprocal yields fractional numbers, these numbers are multiplied up until whole numbers are generated for the Miller indices. The Miller indices for a set of rows whose reciprocal intercept is ($\frac{1}{3}$, $\frac{1}{2}$), for example, is simplified to (2,3) by multiplying by 6.

The situation in three dimensions exactly parallels this argument, and the indices for the lattice planes in *Fig. 4* illustrate this point. Each Miller index, it should be noted, refers not simply to one plane, but to the whole set of parallel planes with these indices. For example, the (111) planes. By convention, these Miller indices are referred to as the $h$, $k$, and $l$ values respectively. Where an index has a negative value, this is represented by a bar over the number, thus an index of (−112) is written as ($\bar{1}$12).

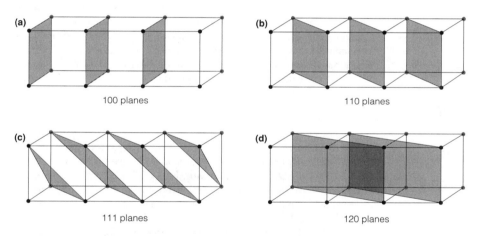

*Fig 4.* Lattice planes in a primitive cubic unit cell.

When considering diffraction methods, it is the distance between planes which becomes the most important consideration. The advantage of the Miller indices is that they enable relatively straightforward calculation of interplane distances. In the simplest situation, that of a cubic crystal of unit cell dimension, $a$, the distance $d_{hkl}$ between planes with Miller indices $h$, $k$, and $l$ is easily calculated from:

$$d_{hkl} = a \, / \, (h^2 + k^2 + l^2)^{1/2}$$

The distance between two (1,2,2) planes in sodium chloride (cubic unit cell, $a = 5.56$ Å) is given by $d_{122} = 5.56 \, / \, (1^2 + 2^2 + 2^2)^{1/2} = 5.56/3 = 1.85$ Å. The corresponding relationships for other unit cell systems, and their relationship to diffraction effects, are discussed in Topic A6.

# A6 DIFFRACTION BY SOLIDS

## Key Notes

**Radiation for diffraction by solids**

Diffraction takes place when a wave interacts with a lattice whose dimensions are of the same order of magnitude as that of the wavelength of the wave. At these dimensions, the lattice scatters the radiation, so as to either enhance the amplitude of the radiation through constructive interference, or to reduce it through destructive interference. The pattern of constructive and destructive interference yields information about molecular and crystal structure. The most commonly used radiation is X-rays, which are most strongly scattered by heavy elements. High velocity electrons behave as waves, and are also scattered by the electron clouds. Neutrons slowed to thermal velocities also behave as waves, but are scattered by atomic nuclei.

**Bragg equation**

In crystallographic studies, the different lattice planes which are present in a crystal are viewed as planes from which the incident radiation can be reflected. Constructive interference of the reflected radiation occurs if the Bragg condition is met: $n\lambda = 2\,d\,\sin\theta$. For most studies, the wavelength of the radiation is fixed, and the angle $\theta$ is varied, allowing the distance between the planes, $d$, to be calculated from the angle at which reflections are observed.

**Reflections**

For a crystalline solid, the distance between the lattice planes is easily obtained from the Miller indices, and the unit cell dimensions. The relationship between these parameters can be used to modify the Bragg condition. In the simple case of a primitive cubic unit cell, the allowed values for $\theta$ as a function of $h$, $k$, and $l$ are given by: $\sin\theta = (h^2 + k^2 + l^2)^{1/2}\,\lambda\,/2a$. Some whole numbers (7, 15, 23, for example) cannot be formed from the sum of three squared numbers, and the reflections corresponding to these values of $(h^2 + k^2 + l^2)$ are missing from the series. In other unit cells, missing lines occur as a result of the symmetry of the unit cell. Simple geometric arguments show that for a body centered cubic unit cell, $h + k + l$ must be even, and that for a face centered cubic unit cell, $h$, $k$ and $l$ must be all even or all odd for reflections to be allowed. The forbidden lines are known as systematic absences.

**Powder crystallography**

In the powder diffraction method, the crystalline sample is ground into a powder, so that it contains crystals which are oriented at every possible angle to the incident beam. In this way, the Bragg condition for every lattice plane is simultaneously fulfilled, and reflections are seen at all possible values of $\theta$. Modern diffractometers use scintillation detectors which sweep an arc of angle $2\theta$ around the sample, giving a measure of X-ray intensity as a function of the angle $2\theta$. The diffraction pattern which is obtained must be correlated with the unit cell of the sample. By obtaining the angles for which reflections occur, the ratios of the values of $\sin^2\theta$ may be directly correlated to the values of $h$, $k$, and $l$ in a process known as indexing.

**Related topics**     Crystalline solids (A5)

**Radiation for diffraction by solids**

**Diffraction** takes place when a wave interacts with a lattice whose dimensions are of the same order of magnitude as that of the wavelength of the wave. The lattice scatters the radiation, and the scattered radiation from one point interferes with the radiation from others so as to either enhance the amplitude of the radiation (**constructive interference**), or to reduce it (**destructive interference**) (*Fig. 1*). The pattern of constructive and destructive interference yields information about molecular and crystal structure.

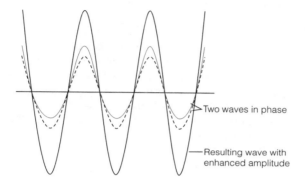

Two waves in phase

Resulting wave with enhanced amplitude

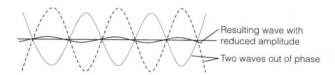

Resulting wave with reduced amplitude

Two waves out of phase

*Fig. 1.   Constructive (a) and destructive (b) interference of two waves.*

In the case of solids, this wavelength must be of the same order as the crystal lattice spacing (ca. 0.1 nm), and there are three primary types of radiation which are used for structural studies of solids. The most commonly used radiation, **X-rays**, have wavelengths of the order of 0.15 nm, and in the course of diffraction studies are scattered by the electron density of the molecule. The heavier elements therefore have the strongest scattering power, and are most easily observed. Similarly, **electrons** which have been accelerated to high velocity may have wavelengths of the order of 0.02 nm, and are also scattered by the electron clouds. Fission-generated **neutrons** which have been slowed to velocities of the order of 1000 m s$^{-1}$ also behave as waves, but are scattered by atomic nuclei. The relationship between scattering power and atomic mass is complex for neutrons. Whilst some light nuclei such as deuterium scatter neutrons strongly, some heavier nuclei, such as vanadium, are almost transparent.

The subjects covered in this topic are indifferent to the nature of the radiation used, and the arguments may be applied to all types of diffraction study.

**Bragg equation**

In crystallographic studies, the different **lattice planes** which are present in a crystal are viewed as planes from which the incident radiation can be reflected.

**Diffraction of the radiation** arises from the phase difference between these reflections. For any two parallel planes, several conditions exist for which constructive interference can occur. If the radiation is incident at an angle, $\theta$, to the planes, then the waves reflected from the lower plane travel a distance equal to $2d \sin\theta$ further than those reflected from the upper plane where $d$ is the separation of the planes. If this difference is equal to a whole number of wavelengths, $n\lambda$, then constructive interference will occur (*Fig. 1*). In this case, the **Bragg condition** for diffraction is met:

$$n\lambda = 2d \sin\theta$$

In all other cases, a phase difference exists between the two beams and they interfere destructively, to varying degrees. The result is that only those reflections which meet the Bragg condition will be observed. In practice, $n$ may be set equal to 1, as higher order reflections merely correspond to first order reflections from other parallel planes which are present in the crystal.

For most studies, the wavelength of the radiation is fixed, and the angle $\theta$ is varied, allowing $d$ to be calculated from the angle at which reflections are observed (*Fig. 2*).

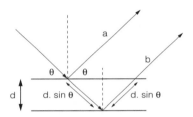

*Fig. 2.   Diffraction due to reflections from a pair of planes. The difference in path length between reflected beams a and b is equal to 2d sinθ. If this is equal to a whole number of wavelengths, nλ, then constructive interference occurs.*

**Reflections**

For a crystalline solid, the distance between the **lattice planes** is easily obtained from the **Miller indices**, and the **unit cell** dimensions. The simplest example is that of a primitive cubic unit cell, for which the distance between planes, $d$, is simply given by:

$$d^2 = a^2 / (h^2 + k^2 + l^2)$$

where $h$, $k$, and $l$ are the Miller indices and $a$ is the length of the unit cell edge. Substitution of this relationship into the Bragg condition yields the possible values for $\theta$:

$$\sin\theta = (h^2 + k^2 + l^2)^{1/2} \lambda / 2a$$

Because $h$, $k$, and $l$ are whole numbers, the sum $(h^2 + k^2 + l^2)$ also yields whole numbers, and because $\lambda$ and $a$ are fixed quantities, $\sin^2\theta$ varies so as to give a regular spacing of reflections. However, some whole numbers (7, 15, 23, etc.) cannot be formed from the sum of three squared numbers, and the reflections corresponding to these values of $(h^2 + k^2 + l^2)$ are missing from the series. If $\lambda/2a$ is denoted $A$, then the values of $\sin\theta$ for a simple cubic lattice are given by: $A/\sqrt{1}$, $A/\sqrt{2}$, $A/\sqrt{3}$, $A/\sqrt{4}$, $A/\sqrt{5}$, $A/\sqrt{6}$, $A/\sqrt{8}$, $A/\sqrt{9}$, $A/\sqrt{10}$, $A/\sqrt{11}$, etc. It is therefore possible to identify a primitive cubic unit cell from both the regularity of the spacings in the X-ray diffraction pattern, and the absence of certain **forbidden lines**.

Other unit cells yield further types of missing lines, known as **systematic absences**. Simple geometric arguments show that the following conditions apply to a cubic unit cell:

|  | Allowed reflections |
|---|---|
| Primitive cubic unit cell | all $h + k + l$ |
| Body centered cubic unit cell | $h + k + l$ = even |
| Face centered cubic unit cell | $h + k + l$ = all even or all odd |

Similar, but increasingly complex, rules apply to other unit cells and identification of the systematic absences allows the unit cell to be classified.

**Powder crystallography**

When a single crystal is illuminated with radiation, **reflections** are only observed when one of the **lattice planes** is at an angle which satisfies the Bragg condition. In the **powder diffraction method**, the crystalline sample is ground into a powder, so that it effectively contains crystals which are oriented at every possible angle to the incident beam. In this way, the Bragg condition for every lattice plane is simultaneously fulfilled, and reflections are seen at all allowable values of $\theta$ relative to the incident beam (*Fig. 3*).

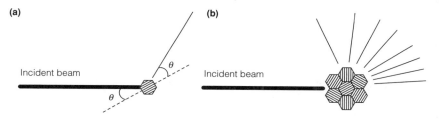

*Fig. 3.* *(a) Diffraction by a single crystal, with one set of lattice planes correctly oriented for an allowed reflection; (b) diffraction by a crystalline powder where some crystals are oriented for every possible allowed reflection.*

Most modern **diffractometers** use scintillation detectors which sweep an arc around the angle $2\theta$. The detector gives a measure of X-ray intensity as a function of the angle $2\theta$. The diffraction pattern which is obtained must be correlated with the unit cell of the sample. By obtaining the angles for which reflections occur, the value of $\sin^2\theta$ may be obtained for each reflection, and these values are directly correlated to the values of $h$, $k$, and $l$:

$$\sin^2\theta = (h^2 + k^2 + l^2)\,(\lambda /2a)^2 \quad \text{(Note that this is simply the square of a previous expression)}$$

As $(\lambda /2a)^2$ is constant, the value of $\sin^2\theta$ is directly related to the value of $(h^2 + k^2 + l^2)$, and for a given crystal the ratios of the values gives the relative values of $(h^2 + k^2 + l^2)$. For example, if X-rays of wavelength 0.1542 nm are incident on a powder sample, and the angular position of the reflections is measured, the process for calculating $(h^2 + k^2 + l^2)$ is given in *Table 1*.

*Table 1. Indexing a simple powder diffraction pattern*

| Angle, $\theta$ | 14.30 | 20.45 | 25.33 | 29.61 | 33.53 | 37.24 | 44.33 | 47.83 |
|---|---|---|---|---|---|---|---|---|
| Calculate $\sin^2\theta$ | 0.061 | 0.122 | 0.183 | 0.244 | 0.305 | 0.366 | 0.488 | 0.549 |
| Ratio | 1 | 2 | 3 | 4 | 5 | 6 | 8 | 9 |

The bottom row of *Table 1* represents the values of $(h^2 + k^2 + l^2)$ corresponding to the reflections at the angles given. The process of ascribing $h$, $k$, and $l$ values to reflections in a diffraction pattern is known as **indexing** the pattern. From the indexed pattern, it is possible to identify the type of unit cell, which in this case can be identified as simple cubic due to the presence of all possible values of $(h^2 + k^2 + l^2)$, with a forbidden line at $(h^2 + k^2 + l^2) = 7$.

# B1 THE FIRST LAW

## Key Notes

**Thermodynamics**

Thermodynamics is the mathematical study of heat and its relationship with mechanical energy and other forms of work. In chemical systems, it allows determination of the feasibility, direction, and equilibrium position of reactions.

**Internal energy**

The sum of all the kinetic and potential energy in a system is the internal energy, $U$. Because it includes nuclear binding energy, and mass–energy equivalence terms, as well as molecular energies, it is not practical to measure an absolute value of $U$. Changes in the value of $U$ and its relationship to other thermodynamic quantities are therefore used.

**State functions and path functions**

If the value of a thermodynamic property is independent of the manner in which it is prepared, and dependent only on the state of that system, that property is referred to as a state function. A path function is a thermodynamic property whose value depends upon the path by which the transition from the initial state to the final state takes place.

**The first law**

The energy of an isolated system is constant. An alternative, equivalent expression is that energy may be neither created nor destroyed, although the energy within a system may change its form. It is a result of the first law that energy in an open system may be exchanged with the surroundings as either work or heat but may not be lost or gained in any other manner.

**Work**

Work is energy in the form of orderly motion, which may, in principle, be harnessed so as to raise a weight. The most common forms of work are pressure–volume work and electrical work. The work done by a system against a constant external pressure is given by $w = -p_{ex}\Delta V$. The maximum amount of volume expansion work which a system may accomplish under reversible conditions is given by $w = -nRT \ln (V_f/V_i)$.

**Heat capacity**

When a system takes up or gives out energy in the form of heat, the temperature change in the system is directly proportional to the amount of heat. At constant pressure,

$$C_v = \left(\frac{\partial V}{\partial T}\right)_v.$$

The heat capacity at constant pressure, $C_p$ and at constant volume, $C_v$, are approximately equal for solids and liquids, but the difference for gases is given by $C_p = C_v + nR$.

**Related topics**

Enthalpy (B2)
Thermochemistry (B3)
Entropy (B4)

Entropy and change (B5)
Free energy (B6)
Statistical thermodynamics (G8)

**Thermodynamics**

**Thermodynamics** is a macroscopic science, and at its most fundamental level, is the study of two physical quantities, energy and entropy. **Energy** may be regarded as the capacity to do work, whilst **entropy** (see Topics B4 and G8) may be regarded as a measure of the disorder of a system. Thermodynamics is particularly concerned with the interconversion of energy as heat and work. In the chemical context, the relationships between these properties may be regarded as the driving forces behind chemical reactions. Since energy is either released or taken in by all chemical and biochemical processes, thermodynamics enables the prediction of whether a reaction may occur or not without need to consider the nature of matter itself. However, there are limitations to the practical scope of thermodynamics which should be borne in mind. Consideration of the energetics of a reaction is only one part of the story. Although hydrogen and oxygen will react to release a great deal of energy under the correct conditions, both gases can coexist indefinitely without reaction. Thermodynamics determines the **potential** for chemical change, not the **rate** of chemical change – that is the domain of chemical kinetics (see Topics F1 to F6). Furthermore, because it is such a common (and confusing) misconception that the potential for change depends upon the release of energy, it should also be noted that it is not energy, but entropy which is the final arbiter of chemical change (see Topic B5).

Thermodynamics considers the relationship between the **system** – the reaction, process or organism under study – and the **surroundings** – the rest of the universe. It is often sufficient to regard the immediate vicinity of the system (such as a water bath, or at worst, the laboratory) as the surroundings.

Several possible arrangements may exist between the system and the surroundings (*Fig. 1*). In an **open system**, matter and energy may be interchanged between the system and the surroundings. In a **closed system**, energy may be exchanged between the surroundings and the system, but the amount of matter in the system remains constant. In an **isolated system**, neither matter nor energy may be exchanged with the surroundings. A system which is held at constant temperature is referred to as **isothermal,** whilst an **adiabatic** system is one in which energy may be transferred as work, but not as heat, i.e. it is thermally insulated from its surroundings. Chemical and biological studies are primarily concerned with closed isothermal systems, since most processes take place at constant temperature, and it is almost always possible to design experiments which prevent loss of matter from the system under study.

Energy is transferred as either heat or work, which, whilst familiar, are not always easily defined. One of the most useful definitions is derived from the mechanical fashion in which energy is transferred as either heat or work. **Heat** is the transfer of energy as disorderly motion as the result of a temperature difference between the system and its surroundings. **Work** is the transfer of energy as orderly motion. In mechanical terms, work is due to energy being expended

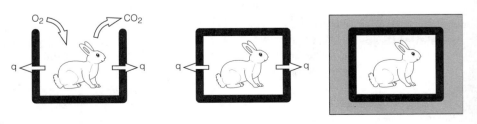

*Fig. 1.   Examples of an open system (left), a closed system (center) and an isolated system (right).*

against an opposing force. The total work is equal to the product of the force and the distance moved against it. Work in chemical or biological systems generally manifests itself in only a limited number of forms. Those most commonly encountered are pressure–volume ($PV$) work and electrical work (see section E).

**Internal energy**

A fundamental parameter in thermodynamics is the **internal energy** denoted as $U$. This is the *total* amount of energy in a system, irrespective of how that energy is stored. Internal energy is the sum total of all kinetic and potential energy within the system. $U$ is a **state function**, as a specific system has a specific value at any given temperature and pressure. In all practical systems, the value of $U$ itself cannot be measured, however, as it involves *all* energy terms including nuclear binding energies and the mass itself. Thermodynamics therefore only deals with changes in $U$, denoted as $\Delta U$. The sign of $\Delta U$ is crucially important. When a system loses energy to the surroundings, $\Delta U$ has a *negative* value, for example, −100 kJ. When the internal energy of a system is increased by gain of energy, $\Delta U$ has a *positive* value, for example +100 kJ. The '+' or '−' sign should always be explicitly written in any thermodynamic calculation, and not simply implied.

**State functions and path functions**

The physical properties of a substance may be classified as extensive or intensive properties. An **extensive property** is one in which the value of the property changes according to the amount of material which is present. The mass of a material is one example, as it changes according to the amount of material present. Doubling the amount of material doubles the mass. The **internal energy** is another example of an extensive property. The value of an **intensive property** is independent of the amount of material present. An example is the temperature or the density of a substance.

An important classification of thermodynamic properties is whether they are **state functions** or **path functions**. If the value of a particular property for a system depends solely on the state of the system at that time, then such a property is referred to as a state function. Examples of state functions are volume, pressure, internal energy and entropy. Where a property depends upon the path by which a system in one state is changed into another state, then that property is referred to as a path function. Work and heat are both examples of path functions. The distinction is important because in performing calculations upon state functions, no account of how the state of interest was prepared is necessary (*Fig. 2*).

**The first law**

The **first law of thermodynamics** states that '*The total energy of an isolated thermodynamic system is constant*'. The law is often referred to as **the conservation of energy**, and implies the popular interpretation of the first law, namely that '*energy cannot be created or destroyed*'. In other words, energy may be lost from a system in only two ways, either as work or as heat. As a result of this, it is possible to describe a change in the total internal energy as the sum of energy lost or gained as work and heat, since $U$ cannot change in any other way. Thus, for a finite change:

$$\Delta U = q + w$$

where $q$ is the heat supplied to the system, and $w$ is the work done on the system. As with $\Delta U$, $q$ and $w$ are *positive* if energy is *gained* by the system as heat

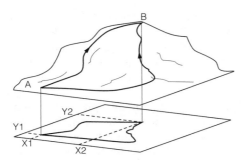

*Fig. 2.   Altitude as a state function. At latitudes and longitudes (X1,Y1) and (X2,Y2) the corresponding altitudes at A and B are fixed quantities, altitude is therefore a state function. The amount of work done and the distance traveled in climbing from A to B depend upon the path. The work and the distance traveled are therefore path functions.*

and work respectively, and *negative* if energy is *lost* from the system as heat or work.

Work and heat are path functions, since the amount of work done or heat lost depends not on the initial and final states of the system, but on how that final state is reached. In changing the internal energy of a system, the amount of energy lost as heat or as work, depends upon how efficiently the energy is extracted. Hence some cars travel further on a given amount of petrol than others depending on how efficiently the internal energy of the petrol is harnessed to do work.

**Work**

There are a limited number of ways in which energy may be exchanged in the form of **work**. The most commonly encountered of these is pressure–volume or *pV* work. Electrical work may also be performed by a system (see Topics E3, E4 and E5), and this may be accounted for by including an appropriate term, but in *most* cases this may be discounted. When a reaction releases a gas at a constant external pressure, $p_{ex}$ work is done in expanding, 'pushing back', the surroundings. In this case, the work done is given by:

$$w = -p_{ex}.\Delta V$$

and so the change in **internal energy $\Delta U$** in such a reaction is:

$$\Delta U = -p_{ex}.\Delta V + q$$

If a reaction is allowed to take place in a sealed container at fixed volume then $\Delta V = 0$, and so the expression for $\Delta U$ reduces to $\Delta U = q$. This is the principle of a **bomb calorimeter**. A bomb calorimeter is a robust metal container in which a reaction takes place (often combustion at high oxygen pressure). As the reaction exchanges heat with the surroundings (a water bath, for example), the temperature of the surroundings changes. Calibration of the bomb using an electrical heater or standard sample allows this temperature rise to be related to the heat output from the reaction and the value of $q$, and hence $\Delta U$, obtained.

**Heat capacity**

When energy is put into a system, there is usually a corresponding rise in the temperature of that system. Assuming that the energy is put in only as **heat**, then the rise in temperature of a system is proportional to the amount of heat which is input into it, and they are related through the **heat capacity**, $C$:

$$dq = C.dT \text{ (infinitesimal change)}$$
$$\text{or } q = C\Delta T \text{ (finite change when } C \text{ is temperature independent)}$$

The heat capacity of a substance depends upon whether the substance is allowed to expend energy in expansion work or not, and hence there are two possible heat capacities, the **constant volume heat capacity**, $C_v$, which is the heat capacity measured at constant volume, and the **constant pressure heat capacity**, $C_p$, which is measured at constant pressure. The two are approximately identical for solids and liquids, but for gases they are quite different as energy is expended in volume expansion work. They are related through the formula:

$$C_p = C_v + nR$$

Since, at constant volume, the heat supplied is equal to the change in internal energy, $\delta U$, it is possible to write:

$$\partial U = C_v \, \partial T \quad \text{or} \quad \Delta U = C_v \Delta T$$

when $C_v$ is independent of temperature.

The **molar heat capacity**, $C_m$ is the heat capacity per mole of substance:

$$C_m = C / n$$

The larger the value of $C_m$ the more heat is required to accomplish a given temperature rise.

# B2 ENTHALPY

## Key Notes

**Enthalpy**

Enthalpy, $H$, is defined by the relationship $H = U + pV$. The enthalpy change, $\Delta H$, for finite changes at constant pressure is given by the expression $\Delta H = \Delta U + p\Delta V$, so making the enthalpy change for a process equal to the heat exchange in a system at constant pressure. For a chemical system which releases or absorbs a gas at constant pressure, the enthalpy change is related to the internal energy change by $\Delta H = \Delta U + \Delta n.RT$, where $\Delta n$ is the molar change in gaseous component.

**Properties of enthalpy**

Enthalpy is a state function whose absolute value cannot be known. $\Delta H$ can be ascertained, either by direct methods, where feasible, or indirectly. An increase in the enthalpy of a system, for which $\Delta H$ is positive, is referred to as an endothermic process. Conversely, loss of heat from a system, for which $\Delta H$ has a negative value, is referred to as an exothermic process. The enthalpy change arising from a temperature change at constant pressure is given by the expression $\Delta H = C_p \Delta T$, providing that $C_p$ does not appreciably change over the temperature range of interest. Where $C_p$ does change, the integral form of the equation, $\Delta H = \int_{T1}^{T2} C_p \, dT$, is used. In a chemical reaction, the enthalpy change is equal to the difference in enthalpy between the reactants and products:

$$\Delta H_{\text{Reaction}} = \Sigma H_{\text{(Products)}} - \Sigma H_{\text{(Reactants)}}.$$

**Kirchhoff's law**

The value of $\Delta H$ for a reaction varies considerably with temperature. Kirchhoff's equation, derived from the properties of enthalpy, quantifies this variation. Where $C_p$ does not appreciably change over the temperature range of interest, it may be expressed in the form

$$\Delta H_{T2} - \Delta H_{T1} = \Delta C_p \Delta T, \text{ or as } \Delta H_{T2} - \Delta H_{T1} = \int_{T1}^{T2} \Delta C_p \, dT \text{ where } \Delta C_p \text{ is a}$$

function of temperature.

**Related topics**

| | |
|---|---|
| The first law (B1) | Entropy and change (B5) |
| Thermochemistry (B3) | Free energy (B6) |
| Entropy (B4) | Statistical thermodynamics (G8) |

## Enthalpy

The majority of chemical reactions, and almost all biochemical processes *in vivo*, are performed under constant pressure conditions and involve small volume changes. When a process takes place under constant pressure, and assuming that no work other than $pV$ work is involved, then the relationship between the heat changes and the internal energy of the system is given by:

$$dU = dq - p_{ex}dV \text{ (infinitesimal change)} \qquad \Delta U = q - p_{ex}\Delta V \text{ (finite change)}$$

The **enthalpy, $H$,** is *defined* by the expression; $H = U + pV$, Hence for a finite change at constant pressure:

$$\Delta H = \Delta U + p_{ex}\Delta V$$

Thus, when the only work done by the system is $pV$ work,

$$\Delta H = q \text{ at constant pressure}$$

Expressed in words, the heat exchanged by a system at constant pressure is equal to the sum of the internal energy change of that system and the work done by the system in expanding against the constant external pressure. The enthalpy change is the heat exchanged by the system under conditions of constant pressure.

For a reaction involving a perfect gas, in which heat is generated or taken up, $\Delta H$ is related to $\Delta U$ by:

$$\Delta H = \Delta U + \Delta n\, RT$$

where $\Delta n$ is the change in the number of moles of gaseous components in the reaction. Hence for the reaction $CaCO_3(s) \longrightarrow CaO(s) + CO_2\ (g)$, $\Delta n = +1$ (1 mole of gaseous $CO_2$ is created), and so $\Delta H = \Delta U + 2.48$ kJ mol$^{-1}$ at 298 K.

**Properties of enthalpy**

The **internal energy**, pressure and volume are all **state functions** (see Topic B1), and since **enthalpy** is a function of these parameters, it too is a state function. As with the internal energy, a system possesses a defined value of enthalpy for any particular system at any specific conditions of temperature and pressure. The absolute value of enthalpy of a system cannot be known, but changes in enthalpy can be measured. Enthalpy changes may result from either physical processes (e.g. heat loss to a colder body) or chemical processes (e.g. heat produced *via* a chemical reaction).

An increase in the enthalpy of a system leads to an increase in its temperature (and *vice versa*), and is referred to as an **endothermic** process. Loss of heat from a system lowers its temperature and is referred to as an **exothermic** process. The sign of $\Delta H$ indicates whether heat is lost or gained. For an exothermic process, where heat is lost *from* the system, $\Delta H$ has a negative value. Conversely, for an endothermic process in which heat is gained *by* the system, $\Delta H$ is positive. This is summarized in *Table 1*. The sign of $\Delta H$ indicates the direction of heat flow and should always be explicitly stated, e.g. $\Delta H = +2.4$ kJ mol$^{-1}$.

*Table 1. Exothermic and endothermic processes*

| Heat change in system | Process | Value of $\Delta H$ |
|---|---|---|
| Heat loss (heat lost to the surroundings) | Exothermic | Negative |
| Heat gain (heat gained from the surroundings) | Endothermic | Positive |

For a system experiencing a temperature change at constant pressure, but not undergoing a chemical change, the definition of the constant temperature heat capacity is used in the form $C_p = (\partial q / \partial T)_p$. Since $\partial q$ equals $\partial H$ at constant pressure, the temperature and enthalpy changes are related through the relationship:

$$\Delta H_{T2-T1} = \int_{T1}^{T2} C_p\, dT$$

where $\Delta H_{T2-T1}$ is the enthalpy difference between temperatures T1 and T2.

Over smaller temperature ranges, within which the value of $C_p$ may be regarded as invariant, this expression simplifies to $\Delta H = C_p\, \Delta T$ at constant pressure.

For chemical reactions, the most basic relationship which is encountered follows directly from the fact that enthalpy is a state function. The enthalpy change which accompanies a chemical reaction is equal to the difference between the enthalpy of the products and that of the reactants:

$$\Delta H_{\text{Reaction}} = \Sigma\, H_{\text{(Products)}} - \Sigma\, H_{\text{(Reactants)}}$$

This form of equation is common to all state functions, and appears frequently within thermodynamics. Similar expressions are found for entropy (see Topics B4 and B5) and free energy (see Topic B6).

**Kirchhoff's law**    Because the **enthalpy** of each reaction component varies with temperature, the value of $\Delta H$ for a chemical reaction is also temperature dependent. The relationship between $\Delta H$ and temperature is given by **Kirchhoff's law** which may be written as

$$\Delta H_{\text{T2}} - \Delta H_{\text{T1}} = \int_{\text{T1}}^{\text{T2}} \Delta C_{\text{p}}\, dT$$

If the change in $C_{\text{p}}$ with temperature is negligible, this expression may be simplified to:

$$\Delta H_{\text{T2}} - \Delta H_{\text{T1}} = \Delta C_{\text{p}}\, \Delta T$$

# B3 THERMOCHEMISTRY

---

## Key Notes

| | |
|---|---|
| **Standard state** | The standard state for a material is defined as being the pure substance at 1 atmosphere pressure, and at a specified temperature. The temperature does not form part of the definition of the standard state, but for historical reasons data are generally quoted for 298 K (25°C). For solutions, the definition of the standard state of a substance is an activity of 1. The standard enthalpy change for a process is denoted as $\Delta H^{\ominus}_{298K}$ with the subscript denoting the temperature. |
| **Biological standard state** | The definition of the biological standard state is identical to the standard state, with the exception of the standard state of hydrogen ion activity, which is defined as equal to $10^{-7}$ or pH = 7. Biological standard conditions are denoted by a superscript '$\oplus$', for example $\Delta H^{\oplus}$. Thermodynamic values for a reaction under standard biochemical conditions only differ from that of the conventional standard state when a proton is lost or gained in that reaction. |
| **Specific enthalpy changes** | For the purposes of concise discussion, the enthalpy changes associated with a number of common generic processes are given specific names, although in thermodynamic terms, these processes are treated identically. |
| **Hess's law** | Hess's law of constant heat summation is primarily a restatement of the first law of thermodynamics. It may be summarized as 'The overall enthalpy change for a reaction is equal to the sum of the enthalpy changes for the individual steps in the reaction measured at the same temperature.' Hess's law is particularly useful in calculating enthalpy changes which cannot be easily measured. |
| **Enthalpy of formation** | Tabulated values of the enthalpy of formation of materials may be used to calculate the enthalpy change associated with a reaction using the following, derived from Hess's law: $$\Delta H_{reaction} = \Sigma \Delta H_f \text{ (products)} - \Sigma \Delta H_f \text{ (reactants)}$$ |
| **Enthalpy of combustion** | The enthalpy of combustion of reactant and product materials may be used to calculate the enthalpy change associated with a reaction in a similar manner to that of the enthalpy of formation: $$\Delta H_{reaction} = \Sigma \Delta H_c \text{ (reactants)} - \Sigma \Delta H_c \text{ (products)}$$ The ease with which $\Delta H_c$ values may be obtained is offset by the more limited scope of the expression. |
| **The Born-Haber cycle** | The Born-Haber cycle is a specific example of Hess's law which allows indirect measurement of the lattice enthalpy for an ionic material from $\Delta H_f$ of the material and the enthalpy changes associated with the formation of gaseous cations and anions from the elements in their standard states. |

| Related topics | The first law (B1) | Entropy and change (B5) |
|---|---|---|
| | Enthalpy (B2) | Free energy (B6) |
| | Entropy (B4) | Statistical thermodynamics (G8) |

**Standard state**

The **enthalpy** changes associated with any reaction are dependent upon the temperature (Topic B2). They are also dependent upon the pressure, and the amounts and states of the reactants and products. For this reason, it is convenient to specify a **standard state** for a substance. The standard state for a substance is defined as being the pure substance at 1 atmosphere pressure, and at a specified temperature. The temperature does not form part of the definition of the standard state, but for historical reasons data are generally quoted for 298 K (25°C). For solutions, the definition of the standard state of a substance is an activity of 1 (see Topic D1).

The definition of a standard state allows us to define **standard enthalpy change** as the enthalpy change when reactants in their standard states are converted into products in their standard states. The enthalpy change may be the result of either a physical or a chemical process. The standard enthalpy change for a process is denoted as $\Delta H^{\ominus}_{298K}$ with the subscript denoting the temperature.

**Biological standard state**

The **standard state** for hydrogen ion concentration is defined as an activity of 1 corresponding to pH = 0. With the exception of, for example, stomach acid, biological systems operate at pH values which are far removed from this highly acidic standard. It is convenient, therefore, for biochemists to define the **biological standard state** of a hydrogen ion solution to be equal to pH = 7, corresponding to an activity of $10^{-7}$. The standard state for all other species is an activity of 1. Biological standard conditions are denoted by a superscript '⊕', for example $\Delta H^{\oplus}$. Thermodynamic values for a reaction at the biological standard state only differ from that of the conventional standard state when a proton is lost or gained in the reaction.

**Specific enthalpy changes**

A number of chemical and physical processes are given specific names in order to aid concise discussion. Thermodynamically, there are no differences between the processes, and the only reason for the use of these specific terms is convenience and brevity. A selection of the more important processes is listed in *Table 1*.

**Hess's law**

Because **enthalpy** is a **state function**, it follows that the absolute enthalpy associated with the reactants and products in a reaction are independent of the process by which they were formed. Consequently, the enthalpy change during the course of a reaction, given by $\Sigma H_{reactants} - \Sigma H_{products}$ is independent of the reaction pathway. **Hess's law of constant heat summation** is a recognition of this fact, and states that:

*'The overall enthalpy change for a reaction is equal to the sum of the enthalpy changes for the individual steps in the reaction measured at the same temperature'.*

The law is particularly useful when measurement of a specific enthalpy change is impractical or unfeasible. This may be illustrated by measurement of the

Table 1.  Definitions of some commonly encountered enthalpy changes

| Quantity | Enthalpy associated with: | Notation | Example |
|---|---|---|---|
| Enthalpy of ionization | Electron loss from a species in the gas phase | $\Delta H_i$ | $Na(g) \longrightarrow Na^+(g) + e^-(g)$ |
| Enthalpy of electron affinity | The gain of an electron by a species in the gas phase | $\Delta H_{ea}$ | $F^-(g) + e^-(g) \longrightarrow F^-(g)$ |
| Enthalpy of vaporization | The vaporization of a substance | $\Delta H_v$ | $H_2O(l) \longrightarrow H_2O(g)$ |
| Enthalpy of sublimation | The sublimation of a substance | $\Delta H_{sub}$ | $CO_2(s) \longrightarrow CO_2(g)$ |
| Enthalpy of reaction | Any specified chemical reaction | $\Delta H$ | $Fe_2O_3 + 3Zn \longrightarrow 2Fe + 3ZnO$ |
| Enthalpy of combustion | Complete combustion of a substance | $\Delta H_c$ | $H_2 + \frac{1}{2}O_2 \longrightarrow H_2O$ |
| Enthalpy of formation | The formation of a substance from its elements in their standard state | $\Delta H_f$ | $2Fe + 3S \longrightarrow Fe_2S_3$ |
| Enthalpy of solution | Dissolution of a substance in a specified quantity of solvent | $\Delta H_{sol}$ | $NaCl(s) \longrightarrow Na^+_{aq} + Cl^-_{aq}$ |
| Enthalpy of solvation | Solvation of gaseous ions | $\Delta H_{solv}$ | $Na^+(g) + Cl^-(g) \rightarrow Na^+_{aq} + Cl^-_{aq}$ |

enthalpy change associated with the burning of carbon to form carbon monoxide. It is practically impossible to prevent formation of some carbon dioxide if the enthalpy change is measured directly. The reaction may be written as either a direct (one-step) or an indirect (two-step) process (*Fig. 1*).

Hess's law indicates that the *total* enthalpy change by either path is identical, in which case $\Delta H_1 = \Delta H_2 + \Delta H_3$, so allowing to be obtained a value for $\Delta H_1$ without the need for direct measurement.

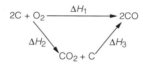

*Fig. 1.   Two possible chemical pathways to the formation of CO from its elements.*

**Enthalpy of formation**

The usefulness of the concept of **enthalpy of formation** (*Table 1*) is readily appreciated when it is used in conjunction with Hess's law. Tables listing the enthalpies of formation of a wide range of materials may be found in the literature, and are more readily available than the enthalpy change associated with a specific reaction. For any reaction, it is possible to construct a reaction pathway which proceeds via the elemental components of both the reactants and the products (*Fig. 2a*). The value for $\Delta H_{reaction}$ is readily calculated from:

$$\Delta H_{reaction} = \Sigma \Delta H_f \text{ (products)} - \Sigma \Delta H_f \text{ (reactants)}$$

Hence for the example reaction in *Fig. 2b*, the reaction enthalpy is given by:

$$\Delta H_{reaction} = [\Delta H_f (CH_3CO_2CH_3) + \Delta H_f (H_2O)] - [\Delta H_f (CH_3CO_2H) + \Delta H_f (CH_3OH)]$$

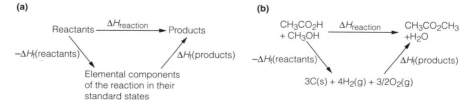

*Fig. 2.   Use of the enthalpy of formation in calculating the enthalpy of a reaction.*

**Enthalpy of combustion**

In the same way as it is possible to usefully combine the enthalpy of formation and Hess's law, it is also possible to combine Hess's law with the **enthalpy of combustion** (Table 1). Taking the previous example, sufficient oxygen may be added to both sides of the equation to formally combust the reactants and products (*Fig. 3*).

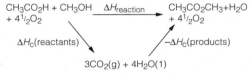

Fig. 3.    Use of the enthalpy of combustion in calculating the enthalpy of a reaction.

The overall enthalpy of reaction is unaffected by this alteration, but $\Delta H_{reaction}$ may now be calculated using Hess's law (note the change of sign as compared to the previous expression):

$$\Delta H_{reaction} = - \Sigma \Delta H_c \text{ (products)} + \Sigma \Delta H_c \text{ (reactants)}$$

The advantage of this method is that enthalpies of combustion are more readily obtained than heats of formation. The disadvantage is that it can only be applied to reactions involving combustible substances, a restriction which generally also excludes materials in solution.

**The Born-Haber cycle**

The **Born-Haber cycle** is a specific application of the **first law of thermodynamics** using **Hess's law**. The cycle allows indirect determination of the **lattice enthalpy** of an ionic solid. This is the enthalpy associated with the direct combination of gaseous ions to form an ionic lattice:

$$nM^{m+}(g) + mX^{n-}(g) \longrightarrow M_nX_m (s)$$

Because direct measurement of this process is generally impractical, an indirect path is created. If the example of KCl is taken, the processes illustrated in *Fig. 4* is obtained.

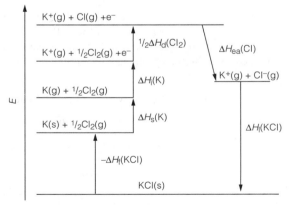

Fig. 4.    The Born-Haber cycle for KCl.

The enthalpy change over the complete cycle must equal zero, since the enthalpy is a state function. Therefore:

$$-\Delta H_f(KCl) + \Delta H_s(K) + \Delta H_i(K) + \tfrac{1}{2}\Delta H_d(Cl_2) + \Delta H_{ea}(Cl) + \Delta H_l(KCl) = 0$$

Rearranging gives:

$$\Delta H_l(KCl) = \Delta H_f(KCl) - \Delta H_s(K) - \Delta H_i(K) - \tfrac{1}{2}\Delta H_d(Cl_2) - \Delta H_{ea}(Cl)$$

The terms on the right hand side of this equation may all be obtained by direct physical or spectroscopic methods, giving a value for the lattice enthalpy: $\Delta H_l(KCl) = -431 - 89 - 419 - 124 - (-349) = -714$ kJ mol$^{-1}$.

# B4 ENTROPY

## Key Notes

**Reversible and irreversible processes**

For any process in which energy is transferred from one body to another, it is possible to transfer the energy in one of two ways. If the energy is transferred reversibly to or from a system, it must be possible to reverse the direction of the transfer through an infinitesimal change in the conditions. In practice, this requires that the energy be transferred infinitely slowly. An irreversible process results from energy transfer which is not transferred under these conditions.

**Thermodynamic definition of entropy**

Entropy is a thermodynamic property of a system, denoted as S. It is a state function and is defined in terms of entropy changes rather than its absolute value. For a reversible process at constant temperature, the change in entropy, $dS$, is given by $dS = dq_{rev}/T$. For an irreversible process, $dS > dq/T$.

**Statistical definition of entropy**

In addition to the thermodynamic definition of entropy, it is also possible to refer to entropy in statistical terms. For any system, the entropy is given by $S = k_B \ln(W)$, where $W$ is the number of possible configurations of the system. This definition allows the entropy to be understood as a measure of the disorder in a system.

**The third law of thermodynamics**

The third law of thermodynamics states that the entropy of a perfectly crystalline solid at the absolute zero of temperature is zero. The entropy has a measurable absolute value for a system, in contrast to the enthalpy and internal energy. There is no requirement for standard entropies of formation to be defined, as the absolute values of entropy may be used in all calculations.

**Related topics**

The first law (B1)
Enthalpy (B2)
Thermochemistry (B3)

Entropy and change (B5)
Free energy (B6)
Statistical thermodynamics (G8)

**Reversible and irreversible processes**

Any process involving the transfer of energy from one body to another may take place either reversibly or irreversibly. In a **reversible** process, energy is transferred in such a way as to ensure that at any point in the process the transfer may be reversed by an infinitesimally small change in the conditions. The system is therefore in equilibrium throughout the transfer. In practice, this means that the energy must be transferred infinitely slowly. An **irreversible** process involves the transfer of energy under any other conditions. In an irreversible process energy is transferred in a manner which results in random motion and some of the energy is dissipated as **heat**. The process is irreversible because a proportion of this heat is dispersed irrecoverably, and the original conditions cannot therefore be generated without **work** being done on the system.

The isothermal expansion of an ideal gas (see Topic A1) against an external pressure is usually given to illustrate the difference between these two conditions. The work, $w$, done by the gas is given by:

$$w = \int_{V1}^{V2} -p.\mathrm{d}V$$

Against a constant pressure (i.e. non-reversible conditions) this integrates to $w = p(V1-V2)$. Under reversible conditions against an infinitesimally smaller pressure, $p$ may be re-written as $(nRT/V)$, and the expression integrates to $nRT\ln(V1/V2)$. The difference is illustrated graphically for one mole of perfect gas expanding from a pressure of 3 bar down to 1 bar in *Fig. 1*. The total amount of work done in each case is equal to the area under the line.

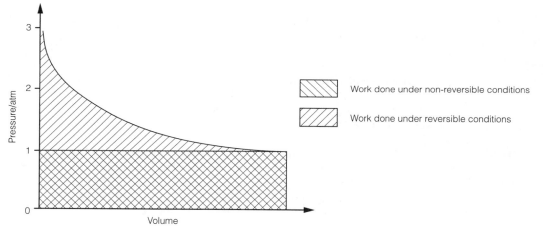

Fig. 1.   *Work done by an expanding gas under reversible and non-reversible conditions.*

**Thermodynamic definition of entropy**

Entropy is a thermodynamic property of a system. It is denoted as S, and like the **enthalpy** and **internal energy**, it is a **state function**. In thermodynamic expressions, entropy is defined in terms of changes in entropy rather than its absolute value. For any process in any system, under isothermal conditions, the change in entropy, dS, is defined as:

$$\mathrm{d}S = \mathrm{d}q_{rev} / T \text{ (reversible process)} \qquad \mathrm{d}S > \mathrm{d}q / T \text{ (irreversible process)}$$

The system entropy change for an irreversible process is unchanged compared to that for a reversible process as entropy is a state function. The entropy change of the surroundings is always $-\mathrm{d}q/\mathrm{d}T$. Thus the *total* entropy change is zero for a reversible process and >0 for an irreversible process. This is the **second law of thermodynamics** (see Topic B5).

It is possible to measure the system entropy changes by measuring the **heat capacity, C**, as a function of temperature. If heat is added reversibly to a system, $\mathrm{d}q_{rev} = C\mathrm{d}T$ and $\mathrm{d}S = C\mathrm{d}T/T$, and the entropy change is then given by:

$$\Delta S = \int_{T1}^{T2} (C/T) \, \mathrm{d}T$$

The area under a plot of $C/T$ against $T$ gives a direct measure of the entropy change in a system (see *Fig. 2*).

For a phase change at constant pressure, $q_{rev}$ is equal to $\Delta H_{phase\ change}$. In the case of fusion, for example, $\Delta S_{fus} = \Delta H_{fus}/T$. In the fusion of 1 mole of mercury at

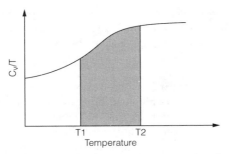

Fig. 2.    Calculation of entropy changes from heat capacity data. The entropy change between T1 and T2 is equal to the shaded area under the curve.

234 K, for example, $\Delta H_{fus}$ = 2333 J, and so, $\Delta S$ = (2333/234) = 9.96 J K$^{-1}$. All phase changes may be similarly treated. The entropy change of vaporization, $\Delta S_{vap} = \Delta H_{vap}/T$, is notable for being dominated by the large absolute entropy of the gas phase. This is very similar for most materials, and gives rise to **Trouton's Rule**, which states that $\Delta S_{vap}$ is approximately equal to 85 J K$^{-1}$ mol$^{-1}$ for most materials. Exceptions to this rule are substances such as water or ammonia, where some degree of ordering in the liquid causes the entropy increase to be greater than this ideal value.

**Statistical definition of entropy**

The **thermodynamic definition of entropy** is constructed in such a way as to aid calculation of entropy changes in real systems. In addition to the thermo-dynamic definition, it is also possible to define entropy in statistical terms, so providing an insight into the real meaning of entropy and entropy changes. For any system, the entropy is given by the **Boltzmann equation**:

$$S = k_B \ln(w)$$

where $w$ is the number of possible configurations of the system and $k_B$ is **Boltzmann's constant**. This definition allows the entropy to be understood as a measure of the disorder in a system. In an example of a hypothetical crystal containing six $^{127}I^{126}I$ molecules, then the number of ways in which the molecules can be arranged if the crystal is perfectly ordered is one (*Fig. 3a*). If two mole-cules are reversed, so increasing the disorder, the number of distinguishable arrangements increases to 15. Reversing three of the molecules further increases the number of possible configurations to 20. If all configurations are energeti-cally equivalent, the most probable arrangement is the one with the highest number of possible configurations, the most 'disordered'. This also means that the perfectly ordered situation, having the lowest number of possible configura-tions and lowest entropy, is the most improbable.

The thermodynamic and statistical descriptions offer different portrayals of entropy, but are both equally valid descriptions of the same concept. The statis-tical definition has the advantage of being conceptually more accessible, but is

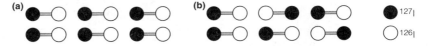

Fig. 3.    (a) The only possible ordered arrangement of six $^{127}I^{126}I$ molecules in a lattice; (b) one of 15 possible arrangements where two of the molecules are reversed.

only practically applicable to very simple and well-defined systems. Although less easily visualized, the thermodynamic definition allows entropy changes to be assessed in complex systems through the use of relatively simple thermo-dynamic measurements.

**The third law of thermodynamics**

The **third law of thermodynamics** states that

> *'the entropy of a perfectly crystalline solid at the absolute zero of temperature is zero'.*

For a perfectly crystalline solid, there can be only one possible spatial configuration of the components of the crystal, and as the material is at the absolute zero of temperature, there are no dynamic changes in the crystal either. Under these conditions, the total number of possible ways of arranging the material equals one ($w = 1$), and so the entropy (defined as $k_B \ln w$) is equal to zero.

Although absolute zero cannot be reached and perfect crystalline solids cannot be made, it is still possible to apply the third law. In practice, the absolute entropy of materials drops to infinitesimally small values at low temperature, and for most purposes equals zero at the low temperatures which can be routinely achieved in the laboratory.

Because it is possible to measure entropy changes from a reference point using heat capacity measurements, entropy (unlike the enthalpy and internal energy) has a measurable absolute value for any system.

# B5 ENTROPY AND CHANGE

---

## Key Notes

**Spontaneous process**

A spontaneous process has a natural tendency to occur without the need for input of work into the system. Examples are the expansion of a gas into a vacuum, a ball rolling down a hill or flow of heat from a hot body to a cold one.

**Non-spontaneous process**

A non-spontaneous process does not have a natural tendency to occur. For a non-spontaneous process to be brought about, energy in the form of work must be put into a system. Examples include the compression of a gas into a smaller volume, the raising of a weight against gravity, or the flow of heat from a cold body to a hotter one in a refrigeration system.

**Second law of thermodynamics**

The second law of thermodynamics states that the entropy of an isolated system increases for irreversible processes and remains constant in the course of reversible processes. The entropy of an isolated system never decreases.

**Standard entropy change**

Because entropy is a state function, entropy changes in a system may be calculated from the standard entropies of the initial and final states of the system:

$$\Delta S^{\ominus}_{process} = S^{\ominus}_{final\ state} - S^{\ominus}_{initial\ state}$$

The standard entropy of reaction is be calculated from:

$$\Delta S^{\ominus} = \Sigma S^{\ominus}(products) - \Sigma S^{\ominus}(products)$$

**Related topics**

The first law (B1)                     Entropy (B6)
Enthalpy (B2)                          Free energy (B6)
Thermochemistry (B3)                   Statistical thermodynamics (G8)

---

**Spontaneous process**

Any process may be defined as being either **spontaneous** or **non-spontaneous**. A **spontaneous process** has a natural tendency to occur, without the need for input of **work** into the system. Examples are the expansion of a gas into a vacuum, a ball rolling down a hill or flow of heat from a hot body to a cold one (*Fig. 1*).

It is important to note that the word 'spontaneous' is a formal definition and is not used in the colloquial sense. If a process is described as spontaneous, it does *not* imply that it is either fast or random. The definition of a spontaneous process has no implications for the rate at which it may come about – a process may be described as spontaneous, but take an infinite amount of time to occur.

For a spontaneous process to take place, the system must be at a position where it is ready for change to come about without the need for work to be done

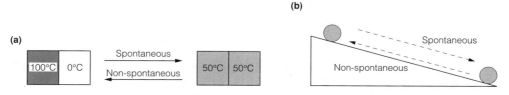

Fig. 1. *Spontaneous and non-spontaneous processes illustrated by heat flow between two bodies in contact (top) and balls on an incline (bottom).*

on it. Indeed, a spontaneous process may be harnessed so as to do work on another system.

**Non-spontaneous process**

A **non-spontaneous** process does not have a natural tendency to occur. Examples include the compression of a gas into a smaller volume, the raising of a weight against gravity, or the flow of heat from a cold body to a hotter one in a refrigeration system.

For a non-spontaneous process to be brought about, energy in the form of **work** must be input into a system. In the case of a ball on a hill, the spontaneous process is for the ball to roll under the influence of gravity to the base of the slope releasing energy as heat in the process. The reverse process – that the ball takes in heat from the surroundings and rolls up the slope – does not occur spontaneously. Note that although the process does not occur naturally, it *is* possible to effect a non-spontaneous process, but now work must be put into the system for this to come about. In the example given, mechanical work must be done in order for the ball to be raised against gravity. In any system, the reverse of a spontaneous process *must* be non-spontaneous.

**Second law of thermodynamics**

The **second law of thermodynamics** is primarily a statement of the probability of the direction in which change proceeds in the universe. It may be stated as

*'The entropy of an isolated system increases for irreversible processes and remains constant in the course of reversible processes. The entropy of an isolated system never decreases'.*

The second law of thermodynamics may be expressed in a large number of ways, but all definitions are equivalent to the one given here. The **statistical definition of entropy** helps visualization of the second law. As all spontaneous changes take place in such a way as to increase the total entropy, it follows that they proceed so as to engineer the chaotic (rather than ordered) dispersal of matter and energy:

$$\Delta S_{total} \geq 0$$

The '>' relation applies to irreversible processes, and the '=' relation applies to reversible processes (see Topic B4). It is important to appreciate that the second law of thermodynamics as expressed above refers to an *isolated* system. Most experimental systems cannot be regarded as being isolated, in which case the universe, being the next largest container of our system, effectively becomes the isolated system. In this case, the total entropy change is simply the sum of the entropy change in the system and in the surroundings, and this total must be greater than or equal to zero to comply with the second law of thermodynamics:

$$\Delta S_{\text{system}} + \Delta S_{\text{surroundings}} = \Delta S_{\text{total}} \geq 0$$

For instance, the system entropy change in the reaction between hydrogen and fluorine gases to generate liquid hydrogen fluoride is found to be $-210$ J K$^{-1}$ mol$^{-1}$. Although this represents a decrease in entropy, the reaction proceeds spontaneously because the total entropy change is greater than zero. The positive entropy change arises because the reaction is exothermic, and the heat lost to the surroundings causes $\Delta S_{\text{surroundings}}$ to be positive, and of greater magnitude than $\Delta S_{\text{system}}$.

**Standard entropy change**

Any non-equilibrium process leads to a change in **entropy**. As entropy is a **state function**, the change may be calculated from the standard entropies of the initial and final states of the system:

$$\Delta S^{\ominus}_{\text{process}} = S^{\ominus}_{\text{final state}} - S^{\ominus}_{\text{initial state}}$$

For a chemical reaction, for example, the **standard entropy of reaction** is therefore the difference between the standard entropies of reactants and products, and may be calculated from:

$$\Delta S^{\ominus}_{\text{reaction}} = \Sigma S^{\ominus}(\text{products}) - \Sigma S^{\ominus}(\text{reactants})$$

This expression resembles those used with other state functions, such as the **enthalpy**, and despite the slightly simpler form, the similarity with expressions for enthalpy is even closer than is initially evident. In the case of enthalpy for example, the corresponding equation is

$$\Delta H^{\ominus}_{\text{reaction}} = \Sigma \Delta H^{\ominus}(\text{products}) - \Sigma \Delta H^{\ominus}(\text{reactants}).$$

Here, $\Delta H_f^{\ominus}$ is the enthalpy of formation of a substance.

# B6 FREE ENERGY

## Key Notes

| | |
|---|---|
| **Free energy** | The Gibbs free energy, $G$, is defined as $G = H-TS$ and at constant pressure and temperature, finite changes in $G$ may be expressed as $\Delta G = \Delta H - T\Delta S$. A similar function, applied at constant volume, is the Helmholtz free energy, $A$, defined as $A = U - TS$. At constant temperature, $\Delta G$ is equal to $-T\Delta S_{total}$ at constant pressure and $\Delta A$ is equal to $-T\Delta S_{total}$ at constant volume. For a spontaneous process $\Delta G < 0$ (constant pressure), or $\Delta A < 0$ (constant volume). Because most chemical and biochemical systems operate at constant pressure, the Gibbs free energy is more commonly encountered. |
| **General properties of the free energies** | The Gibbs and Helmholtz free energies are state functions which do not have measurable absolute values. The free energy change represents the maximum amount of work, other than volume expansion work, which may be obtained from a process. |
| **Free energy and spontaneity** | $\Delta G$ is negative for a spontaneous process. An exothermic reaction ($\Delta H > 0$) with a positive entropy ($\Delta S > 0$) is always spontaneous. A reaction for which $\Delta H < 0$ and $\Delta S < 0$ is spontaneous only at low temperatures, whilst a reaction for which $\Delta H > 0$ and $\Delta S > 0$ is spontaneous only at high temperatures. The temperature at which the reaction becomes spontaneous in each case is given by $T = \Delta H/\Delta S$. |
| **Temperature dependence of the Gibbs free energy** | The most useful expression for the temperature dependence of the Gibbs free energy is the Gibbs-Helmholtz equation: $$\left( \frac{\partial}{\partial T} \left( \frac{G}{T} \right) \right)_p = \frac{-H}{T^2}$$ The Gibbs–Helmholtz expression is most useful when applied to changes in G at constant pressure, such as in the course of a chemical reaction, when it may be written in the form: $$\left( \frac{\partial}{\partial T} \left( \frac{\Delta G}{T} \right) \right)_p = \frac{-\Delta H}{T^2}$$ |
| **Properties and applications of the Gibbs free energy** | The standard reaction free energy, $\Delta G^{\ominus}$, is the change in the Gibbs free energy which accompanies the conversion of reactants in their standard states into products in their standard states. It may be calculated from the enthalpy and entropy changes for a reaction using $\Delta G^{\ominus} = \Delta H^{\ominus} - T\Delta S^{\ominus}$, or from tabulated values for the standard free energy of formation, $\Delta G_f^{\ominus}$, for the reaction components using: $$\Delta G^{\ominus}_{reaction} = \sum \Delta G_f^{\ominus}(\text{products}) - \sum \Delta G_f^{\ominus}(\text{reactants})$$ A more general concept, the reaction free energy, is the change in free energy when a reaction takes place under conditions of constant |

composition. The reaction free energy varies markedly with composition through the expression $\Delta G = \Delta G^{\ominus} + RT \ln Q$ where $Q$ is the reaction quotient.

**Related topics**        The first law (B1)                    Entropy (B4)
                          Enthalpy (B2)                        Entropy and change (B5)
                          Thermochemistry (B3)

**Free energy**           The total **entropy** change which accompanies a process is the sum of the entropy change in the system and its surroundings:

$$\Delta S_{total} = \Delta S_{system} + \Delta S_{surroundings}$$

$\Delta S_{surroundings}$ is related to the **enthalpy** change in the system at constant pressure through the relationship: $\Delta S_{surroundings} = -\Delta H_{system}/T$. Substitution of this expression into the previous one, and subsequent multiplication by $-T$ yields the relationship:

$$-T\Delta S_{total} = \Delta H_{system} - T\Delta S_{system}$$

The **Gibbs free energy, $G$,** (occasionally referred to as the **Gibbs energy** or **Gibbs function**) is defined by $G = H - TS$. At constant pressure and temperature, finite changes may be expressed as:

$$\Delta G = \Delta H_{system} - T\Delta S_{system}$$

$\Delta G$ is therefore equal to $-T\Delta S_{total}$, and the free energy may be regarded as a measure of the total entropy change (both in the system and the surroundings) for a process. Whilst a **spontaneous process** gives rise to a positive value of $\Delta S$, $\Delta G$ must be negative because of the minus sign in $\Delta G = -T\Delta S_{total}$.

$$\Delta G < 0 \qquad \text{for a spontaneous process at constant pressure}$$

A similar function, used for work at constant volume and temperature, is termed the **Helmholtz free energy,** $A$ (also known as the **Helmholtz energy** or **Helmholtz function**). As $\Delta S_{surroundings} = -\Delta U_{system}/T$, under these conditions, the Helmholtz free energy is defined as $A = U - TS$, and $\Delta A$ is therefore equal to $-T\Delta S_{total}$ at constant volume.

$$\Delta A < 0 \qquad \text{for a spontaneous process at constant volume}$$

The Helmholtz free energy is useful in **closed systems** where changes occur (or may be approximated to occur) under constant volume conditions, such as reactions or processes in solids. However, because most chemical and biochemical systems take place at constant pressure, the Gibbs free energy is by far the more commonly encountered property.

**General properties of the free energies**        Because they are wholly derived from **state functions**, it follows that both the Gibbs and Helmholtz free energies are also state functions. Both of the free energies do not have measurable absolute values, and calculations involving free energy changes may be manipulated in the same manner as, for example, enthalpy changes. Hess's law may be applied to free energies, and it is similarly useful to define free energies of formation for substances.

A most important property of the free energy is that it not only provides an indication of the spontaneity of a process but it also represents the maximum amount of work, other than volume expansion work, which may be obtained from a process. This differs from the heat which may be obtained from a process, because the total entropy change must be greater than zero. For example, in the case of a reaction for which $\Delta S_{system}$ is negative, some heat must be lost to the surroundings and contribute to $\Delta S_{surroundings}$ in order that $\Delta S_{total}$ is greater than zero. The value of the heat which is then unavailable for conversion into work is given by $T\Delta S_{system}$.

**Free energy and spontaneity**

For a **spontaneous process**, $\Delta S_{total}$ is positive and $\Delta G$ is therefore negative. The relationship $\Delta G = \Delta H - T\Delta S_{system}$ allows prediction of the conditions under which a reaction is spontaneous. As $T$ must be positive, the relationships may be summarized in *Table 1*.

Table 1.  Free energy and the spontaneity of reactions

| $\Delta H$ | $\Delta S$ | Spontaneous? | Spontaneity favored by |
|------------|------------|--------------|------------------------|
| Negative | Positive | Under all conditions | All conditions |
| Negative | Negative | If $|T\Delta S| < |\Delta H|$ | Low temperatures |
| Positive | Positive | If $|T\Delta S| > |\Delta H|$ | High temperatures |
| Positive | Negative | Never | No conditions |

Temperature has a major impact on the spontaneity of some reactions as indicated in *Table 1*. For a reaction where $\Delta H < 0$ and $\Delta S < 0$, $|T\Delta S|$ will be less than $|\Delta H|$ provided that $T$ is small, and such a reaction will be spontaneous at lower temperatures. Conversely, when $\Delta H > 0$ and $\Delta S > 0$, $|T\Delta S|$ will be greater than $|\Delta H|$ provided that $T$ is large, and such a reaction will become spontaneous at higher temperatures. In both cases, the temperature at which the reaction becomes spontaneous (when $\Delta G = 0$) is simply given by $T = \Delta H / \Delta S$.

**Temperature dependence of the Gibbs free energy**

For a closed system doing no work other than that due to volume expansion, it is possible to show that $dG = Vdp - SdT$ and so it follows that, at constant pressure,

$$\left(\frac{\partial G}{\partial T}\right)_p = -S$$

which provides one relationship between the **Gibbs free energy** and temperature. However, it is possible to take this further, and in doing so obtain a more useful relationship. Since, by definition, $G = H-TS$, then $-S = (G-H)/T$, which may be substituted in the previous expression and the result rearranged to give:

$$\left(\frac{\partial G}{\partial T}\right)_p - \frac{G}{T} = \frac{-H}{T}$$

The left hand side of this expression simplifies to $T\left(\frac{\partial}{\partial T}\left(\frac{G}{T}\right)\right)_p$ to give the

**Gibbs-Helmholtz equation**:

$$\left(\frac{\partial}{\partial T}\left(\frac{G}{T}\right)\right)_p = \frac{-H}{T^2}$$

This expression is most useful when applied to changes in $G$ at constant pressure, such as in the course of a chemical reaction, when it may be written in the form:

$$\left(\frac{\partial}{\partial T}\left(\frac{\Delta G}{T}\right)\right)_p = \frac{-\Delta H}{T^2}$$

**Properties and applications of the Gibbs free energy**

The Gibbs free energy can be applied in a similar manner to other state functions, and many of the expressions which are encountered are similar in form to those seen for the enthalpy, for example (see Topic B2).

The **standard reaction free energy**, $\Delta G^\ominus$, is the change in the Gibbs free energy which accompanies the conversion of reactants in their standard states into products in their standard states. It is possible to calculate the free energy of a reaction from the standard enthalpy and energy changes for the reaction: $\Delta G^\ominus = \Delta H^\ominus - T\Delta S^\ominus$, with $\Delta H^\ominus$ and $\Delta S^\ominus$ being obtained either from tabulated data or direct measurement. An alternative is to use the **standard free energy of formation, $\Delta G_f^\ominus$**. This is defined as the free energy which accompanies the formation of a substance in its standard state from its elements in their standard states. Calculation of the standard free energy of a reaction may be expressed as:

$$\Delta G^\ominus_{reaction} = \sum \Delta G_f^\ominus (\text{products}) - \sum \Delta G_f^\ominus (\text{reactants})$$

$\Delta G_f^\ominus$ values may be obtained from standard tables. Substances with negative values of $\Delta G_f^\ominus$ are termed **thermodynamically stable**. Substances which have positive values of $\Delta G_f^\ominus$ are termed **thermodynamically unstable**. Thermodynamically unstable materials may be synthesized under non-standard conditions and remain stable due to kinetic factors, but cannot be formed directly from the elements in their standard states.

The standard reaction free energy is limited in its usefulness, as it requires that both reactants and products be in their standard states. The **reaction free energy** is the change in free energy when a reaction takes place under conditions of constant composition. The difference may be illustrated by the reaction:

cyclopropane, $C_3H_6$ (g) $\leftrightarrows$ propene, $C_3H_6$ (g)         $\Delta G^\ominus_{298} = -41.7$ kJ mol$^{-1}$

If, rather than measuring the standard free energy change, the free energy change is measured when the molar ratio of the reaction components is 1:2, it is found to be $-39.9$ kJ mol$^{-1}$ at 298 K. The difference arises because of the different conditions which the two values relate to.

The reaction free energy varies markedly with composition, and is directly related to the standard reaction free energy through the **reaction quotient, $Q$** (see Topic C1):

$$\Delta G = \Delta G^\ominus + RT \ln Q$$

# C1 FUNDAMENTALS OF EQUILIBRIA

## Key Notes

| | |
|---|---|
| **Conditions of equilibrium** | An equilibrium is established when the rate of the forward and backward reactions are equal. This is denoted for a general reaction by:<br><br>$$aA + bB \rightleftharpoons cC + dD$$<br><br>At equilibrium, the Gibbs free energy changes for both the forward and backward reactions are zero. The equilibrium constant, $K$, is given by:<br><br>$$K = \frac{a_C^c a_D^d}{a_A^a a_B^b}$$<br><br>where $a_i$ is the activity of species $i$. |
| **$K$ in concentrations** | For systems with negligible interaction between the reacting species (gases, neutral molecules and ions at low dilution) the equilibrium constant is given by:<br><br>$$K = \left(\frac{c_C^c c_D^d}{c_A^a c_B^b}\right)(c^\ominus)^{a+b-c-d}$$<br><br>where $c_i$ is the concentration of species $i$ and $c^\ominus$, the standard concentration, is 1 mol dm$^{-3}$. The equilibrium constant can therefore be calculated from the concentrations of the reacting species in mol dm$^{-3}$. |
| **Gases, solids and pure liquids** | The equilibrium constant for the general reaction with all species in the gaseous state is:<br><br>$$K = \left(\frac{p_C^c p_D^d}{p_A^a p_B^b}\right)(p^\ominus)^{(a+b-c-d)}$$<br><br>where $p_i$ is the partial pressure of the gaseous species $i$ and $p^\ominus$ is the standard pressure of 1 atmosphere. The equilibrium constant can therefore be calculated from the partial pressures of the reacting species in atmospheres. Pure solids and pure liquids have unit activity; this means there are no partial pressure or concentration terms for pure solid or pure liquid species in the equilibrium constant expression. |
| **Physical transitions** | An equilibrium can also exist between the same species in two different forms e.g. solid and liquid, liquid and gas. An equilibrium constant may also be produced for each of these systems. |
| **Thermodynamic data from $K$** | The standard Gibbs free energy for the forward reaction is related to the equilibrium constant by the expression:<br><br>$$\Delta G^\ominus = -RT\ln K$$ |

The standard enthalpy change for the forward reaction can be obtained from the variation of ln$K$ with temperature:

$$\left(\frac{\mathrm{d}\ln K}{\mathrm{d}T}\right) = \frac{\Delta H^{\ominus}}{RT^2}$$

The standard entropy change is then calculable as $\Delta G^{\ominus} = \Delta H^{\ominus} - T\Delta S^{\ominus}$.

**Response to changes**

The equilibrium position changes to oppose any perturbation to the system, in accordance with Le Chatelier's principle. Thus increasing the pressure causes the equilibrium to shift to reduce the overall pressure rise. Increasing the temperature causes the equilibrium position to move in order to reduce the temperature rise. Adding a reactant or a product causes a change in equilibrium position that removes this species.

**Related topics**

Thermochemistry (B3)             Ions in aqueous solution (E1)
Free energy (B6)                 Thermodynamics of ions in solution
Non-electrolyte solutions (D1)    (E2)

**Conditions of equilibrium**

For the general reaction:

$$a\mathrm{A} + b\mathrm{B} \rightarrow c\mathrm{C} + d\mathrm{D}$$

the change in **Gibbs free energy**, $\Delta G$ (see Topic B6) for the reaction at a temperature, $T$, is given by

$$\Delta G = \Delta G^{\ominus} + RT\ln Q$$

where $Q$ is the **reaction quotient**,

$$Q = \frac{a_{\mathrm{C}}^{c} a_{\mathrm{D}}^{d}}{a_{\mathrm{A}}^{a} a_{\mathrm{B}}^{b}}$$

and $a_i$ is the **activity** of species $i$ (see Topic D1). $\Delta G^{\ominus}$ is the standard free energy change for the reaction, defined as the free energy change when all the **reactants** (or **reagents**, i.e. A, B) and **products** (i.e. C, D) have unit activity. When $\Delta G$ is negative, the reaction is **spontaneous** (see Topic B6) and will occur in the direction shown. Conversely, when $\Delta G$ is positive, the reaction is not spontaneous, but the reverse reaction, $c\mathrm{C} + d\mathrm{D} \rightarrow a\mathrm{A} + b\mathrm{B}$, is, as is its free energy change, $\Delta G_{\mathrm{back}} = -\Delta G$.

Under one particular condition, the free energy change of both the forward and the reverse reaction is equal. This is when $\Delta G_{\mathrm{back}} = \Delta G = 0$. In this case, the **rate constants** for the forward and back reactions, $k_f$ and $k_b$, and the activities (and the concentrations, see Topic D1) of reactants and products are such that both forward and reverse reactions are occurring at an equal **rate** (see Topic F1):

$$a\mathrm{A} + b\mathrm{B} \underset{k_b}{\overset{k_f}{\rightleftarrows}} c\mathrm{C} + d\mathrm{D}$$

and the reaction is at **equilibrium**. At equilibrium, the system is at **steady-state** (i.e. the activities of reactants and products remain unchanged), but it should be emphasized this does not mean that no reaction is occurring. Rather, as the rates of the forward and backward reactions are equal, the disappearance of reactants due to the forward reaction is exactly balanced by their appearance due to the backward reaction. This dynamic condition is usually emphasized by using half-headed arrows, i.e.

$$aA + bB \rightleftharpoons cC + dD$$

From the equation above, at equilibrium:

$$\Delta G = 0 = \Delta G^\ominus + RT\ln K$$

where $K$ is the particular value of $Q$ at equilibrium. $K$ is called the **equilibrium constant** for the forward reaction and is given by:

$$K = \frac{a_C^c a_D^d}{a_A^a a_B^b}$$

where the activity values are such that equilibrium is established.

The magnitude of $K$ is also given by the ratio of the forward to backward rate constants (see Topic F5), $K = \dfrac{k_f}{k_b}$, which further emphasizes the dynamic nature of the process.

From this equation:

$$\Delta G^\ominus = - RT\ln K$$

which allows the standard free energy change for this forward reaction (with its **stoichiometries** $a, b, c, d$) to be derived from the equilibrium constant.

**K in concentrations**

The **activity**, $a_i$, given by:

$$a_i = \frac{\gamma_i c_i}{c^\ominus}$$

where $c_i$ is the concentration of a species, $i$, in mol dm$^{-3}$ (see Topic D1), $c^\ominus$ is the standard concentration of 1 mol dm$^{-3}$ and $\gamma_i$ is the **activity coefficient** of the species. The activity therefore depends on the species concentration, but this is modified by the **activity coefficient** $\gamma_i$, which takes account of the (usually attractive) interactions between species, which can stabilize them by reducing their free energy. The activity coefficient itself depends on concentration, as at a higher concentration, species are closer together and tend to interact more. This is an extra complication when determining an equilibrium constant from the concentrations of reagents and products. In many cases this complication can be avoided, as for neutral molecules, both in the gas phase and in solution, the intermolecular forces are relatively weak and short-range, and can be neglected at all practical concentrations. This is also the case for ions (which are charged chemical species, see Topic E1) at low solution concentrations (typically of the order of 10$^{-4}$ mol dm$^{-3}$ and below), when the ions are sufficiently separated in solution that the electrostatic interactions between them can be neglected (see Topics E1 and E2).

Under these conditions of negligible interaction, $\gamma_i$ can be assumed to be unity for all species and by substituting concentration for activities, the equilibrium constant for the reaction:

$$aA + bB \underset{k_b}{\overset{k_f}{\rightleftharpoons}} cC + dD \qquad \text{becomes}$$

$$K = \left( \frac{c_C^c c_D^d}{c_A^a c_B^b} \right)(c^\ominus)^{(a+b-c-d)}$$

where $c_i$ is the **concentration** of species $i$ (see Topic D1). In this general case, the value of the equilibrium constant for the reaction is therefore simply found by combining the concentrations of the reactants and products (in units of mol dm$^{-3}$) at equilibrium, producing a dimensionless equilibrium constant.

However, the activity coefficients cannot be approximated to unity for ions at higher concentrations, as the electrostatic forces between them become important (see Topics E1 to E4 and C5).

**Gases, solids and pure liquids**

When the equilibrium reaction involves gases, it is more convenient to measure the amount of these species by their **partial pressures** (see Topic D2). The activity of a gas, $i$ (often called its **fugacity**) is given by:

$$a_i = \gamma_i \frac{p_i}{p^{\ominus}} \approx \frac{p_i}{p^{\ominus}}$$

where $p_i$ is the partial pressure of $i$. The approximation can be applied as gas molecules are neutral and widely separated, which means that $\gamma_i \approx 1$. The **standard pressure** for a gas, $p^{\ominus}$, is defined as 1 atmosphere. Thus, for an equilibrium involving gases, such as $aA_{(g)} + bB_{(g)} \underset{k_b}{\overset{k_f}{\rightleftharpoons}} cC_{(g)} + dD_{(g)}$, the equilibrium constant is given by:

$$K = \left( \frac{p_C^c p_D^d}{p_A^a p_B^b} \right)(p^{\ominus})^{(a+b-c-d)}$$

and, in a similar fashion to concentrations, the values of the partial pressures of the gases at equilibrium (in units of atmospheres, see Topic D1) can be used to obtain the dimensionless value of $K$. For equilibria involving gases and species in solution, for example the solubility of a gas in a liquid, the partial pressures and concentration terms of the respective species, as appropriate, are combined to produce the equilibrium constant expression.

When a **pure solid** or a **pure liquid** is present in the equilibrium, such as:

$$aA_{(s)} + bB_{(g)} \underset{k_b}{\overset{k_f}{\rightleftharpoons}} cC_{(g)} + dD_{(g)}$$

it is clearly meaningless to talk of a variation in concentration within a pure substance such as A (see Topic D1). For these substances the activity is unity under all conditions. This is equivalent to ignoring pure solids and liquids when drawing up the equilibrium expression, so that no terms (activity, $a$, concentration, $c$, or pressure, $p$, as appropriate) due to these species are present, which in the above example gives:

$$K = \left( \frac{p_C^c p_D^d}{p_B^b} \right)(p^{\ominus})^{(b-c-d)}$$

**Physical transitions**

An equilibrium can also exist between the same species in two different forms e.g. solid and liquid, liquid and gas. This is a **physical transition** or a transition in the **state** of the system. For an equilibrium between a liquid A and a gas A (the vaporization and condensation of A):

$$A_{(l)} \rightleftharpoons A_{(g)}$$

the equilibrium constant is given by the expression:

$$K = \frac{p_A}{p^\ominus}$$

since A is a pure liquid. Thus the vapor pressure of the gas, $p_A$ (in atmospheres), gives the value of the equilibrium constant directly. This is also the case for a system which **sublimes** (transforms directly from a pure solid to a gas). For a system where a species is soluble in two immiscible liquid phases, i.e. $A_{(l1)} \rightleftharpoons A_{(l2)}$ where A **partitions** between liquid 1 (l1) and liquid 2 (l2), the equilibrium constant for the partitioning process is:

$$K = \frac{a_{A,l2}}{a_{A,l1}} \approx \frac{c_{A,l2}}{c_{A,l1}}$$

where $c_{A,l2}$ and $c_{A,l1}$ are the concentrations of A in the liquids l2 and l1 respectively if $\gamma_{A,l1} = \gamma_{A,l2}$. This equilibrium constant is termed a **partition coefficient**.

**Thermodynamic data from K**

Given that (see Topic B6).

$$\Delta G^\ominus = \Delta H^\ominus - T\Delta S^\ominus = -RT\ln K$$

then $\Delta S^\ominus$ values can be determined at any T if $\Delta H^\ominus$ can be obtained. Combining these expressions, $\ln K = \dfrac{-\Delta G^\ominus}{RT} = \dfrac{\Delta S^\ominus}{R} - \dfrac{\Delta H^\ominus}{RT}$, and for a small change in temperature, d$T$, both $\Delta S^\ominus$ and $\Delta H^\ominus$ can be assumed to be independent of $T$. Therefore at constant pressure:

$$\left( \frac{\mathrm{d}\ln K}{\mathrm{d}T} \right) = \frac{\Delta H^\ominus}{RT^2}$$

This equation is called the **van't Hoff equation** and allows $\Delta H^\ominus$ to be measured at any temperature, $T$. This is done by using the small changes in the concentrations and pressures of the species in equilibrium to calculate the small change in $\ln K$, d$\ln K$, when $T$ is changed by a small amount d$T$ from $T$ to $(T + \mathrm{d}T)$.

Often, $\Delta H^\ominus$ and $\Delta S^\ominus$ show little variation over a much wider temperature range. In this case the equation:

$$\ln K_{T1} - \ln K_{T2} = \ln\left( \frac{K_{T1}}{K_{T2}} \right) = \frac{-\Delta H^\ominus}{R}\left[ \frac{1}{T_1} - \frac{1}{T_2} \right]$$

can be used to determine $\Delta H^\ominus$, where $K_{T1}$ and $K_{T2}$ are the equilibrium constants measured at any two temperatures, $T_1$ and $T_2$ respectively, within this range.

A special example of these equations is the equilibrium produced by the physical transition of a species from a liquid to a gas (**vaporization**) and the reverse reaction, where the gas becomes a liquid (**condensation**):

$$A_{(l)} \rightleftharpoons A_{(g)}$$

In this case

$$K = \frac{p_A}{p^\ominus}$$

and so

$$\left( \frac{\mathrm{d}\ln\left(\frac{p}{p^{\ominus}}\right)}{\mathrm{d}T} \right) = \frac{\Delta H_{\mathrm{vap}}^{\ominus}}{RT^2}$$

where $\Delta H^{\ominus}_{\mathrm{vap}}$ is the **standard enthalpy of vaporization** of A (see Topic B3). This expression is called the **Clausius-Clapeyron equation**. The equivalent expression for a temperature range where $\Delta H^{\ominus}_{\mathrm{vap}}$ can be assumed to be constant is:

$$\ln\left(\frac{p_1}{p_2}\right) = \frac{-\Delta H_{\mathrm{vap}}^{\ominus}}{R}\left[\frac{1}{T_1} - \frac{1}{T_2}\right]$$

which allows $\Delta H^{\ominus}_{\mathrm{vap}}$ to be determined from two measurements of the **gas vapor pressure**, $p_1$ and $p_2$, each at their respective temperature $T_1$ and $T_2$.

**Response to changes**

For the general reaction $a\mathrm{A} + b\mathrm{B} \rightleftharpoons c\mathrm{C} + d\mathrm{D}$ at equilibrium, $\Delta G = 0$ and

$$K = \left(\frac{c_{\mathrm{C}}^c c_{\mathrm{D}}^d}{c_{\mathrm{A}}^a c_{\mathrm{B}}^b}\right)(c^{\ominus})^{(a+b-c-d)}$$

However, when a species on the left-hand side (LHS) of the equation, e.g. A, is added so that $c_{\mathrm{A}}$ increases, this removes the equilibrium condition, decreases **the reaction quotient** $Q$ and hence makes $\Delta G$ negative. The forward reaction becomes spontaneous and dominates the backward reaction, and species on the LHS of the equation (A, B) are consumed in order to produce more on the RHS (C, D). This continues until a new equilibrium position is reached, for which the equations $\Delta G = 0$ and hence $\Delta G^{\ominus} = -RT\ln K$ again apply and the concentrations are again related by the equilibrium constant expression above and the value of $K$ remains unchanged. In contrast, if C or D is added, this again perturbs the equilibrium, $Q$ increases, $\Delta G$ becomes positive and the backward reaction is favored over the forward reaction. Equilibrium is again re-established with the consumption of C, D and the production of A, B until the concentrations are related by the equation for $K$ above, with the value of $K$ remaining unchanged. A further perturbation to the system could be to increase the overall pressure of a system involving gases. For example, for:

$$2\mathrm{NH}_{3(g)} \rightleftharpoons 2\mathrm{N}_{2(g)} + 3\mathrm{H}_{2(g)}$$

the equilibrium constant for the reaction as written is given by:

$$K = \frac{p_{\mathrm{N}_2}^2 p_{\mathrm{H}_2}^3}{p_{\mathrm{NH}_3}^2 (p^{\ominus})^3}$$

Increasing the overall pressure causes an increase in all of the partial pressures. As the equilibrium constant involves more moles of gas on the RHS of the equation than the LHS, equilibrium is lost and the reaction quotient, $Q$, becomes larger than $K$. From the equation:

$$\Delta G = \Delta G^{\ominus} + RT\ln Q$$

$\Delta G$ becomes positive, the backward reaction becomes spontaneous and $\mathrm{N}_2$ and $\mathrm{H}_2$ react to form $\mathrm{NH}_3$ until the partial pressures are again related by the equilibrium constant expression given above. In contrast, if more moles of gas were present overall on the LHS of an equilibrium compared with the RHS, $Q$ would decrease when the overall pressure was increased, $\Delta G$ would become negative, the forward reaction would become spontaneous and would occur, decreasing the amount of gas in the system. Equilibrium would again be re-established when the equilibrium partial pressures were related by the equilibrium constant expression, with an overall decrease in pressure.

The system could also be perturbed by a change in temperature rather than concentration. In this case, the equilibrium constant would change value according to the expression $(\mathrm{d}\ln K/\mathrm{d}T) = \Delta H^{\ominus}/RT^2$, so that if $\Delta H^{\ominus}$, the change in enthalpy of the forward reaction were **endothermic** (see Topic B3) an increase in the temperature, $T$, of the system would increase the value of the equilibrium constant K. The equilibrium condition would therefore be lost and the existing reaction quotient, $Q$ (the value of the old equilibrium constant, K) would be less than this new equilibrium constant, and $\Delta G$ for the forward reaction would be negative. This means that C, D would be produced at the expense of A, B until equilibrium were re-established. This would be an **endothermic** process, with heat being taken up from the system during the reaction, reducing the initial temperature rise.

In contrast, if the forward reaction were **exothermic** (see Topic B3), increasing $T$ would decrease the value of K, making $\Delta G$ positive for the forward reaction (the reverse reaction spontaneous) and A, B would be produced at the expense of C, D until equilibrium were re-established with the concentrations related by the value of the new equilibrium constant. Again this process would be endothermic, and heat would be taken up during the reaction, which would again reduce the initial temperature rise.

These are all examples of **Le Chatelier's Principle**, which states that:

*'when a system at equilibrium is subjected to a disturbance, the composition adjusts to minimize the effect of this disturbance.'*

Thus, when a chemical species that forms part of the equilibrium reaction is added to the system at equilibrium, reaction occurs to remove that species. Also when the total pressure of a system involving gases at equilibrium is increased, the system adjusts to reduce the total number of moles of gas (and hence the volume) and offset this pressure increase. Finally, when the temperature of a system is increased, the system adjusts to take in energy and reduce this temperature increase. This is a useful principle that allows the effect of any perturbation on the equilibrium to be predicted.

# C2 FUNDAMENTALS OF ACIDS AND BASES

## Key Notes

| | |
|---|---|
| **Brønsted-Lowry theory** | A Brønsted acid is a proton donor, whilst a Brønsted base is a proton acceptor. An acid–base reaction involves the exchange of a proton between an acid and a base. An acid reacts with water to produce the hydronium ion, $H_3O^+$, and a conjugate base. A base reacts with water to produce the hydroxide ion, $OH^-$, and a conjugate acid. |
| **The autoprotolysis constant** | Water can act as both an acid and a base. Pure water contains both hydronium and hydroxide ions, and their activities are linked by the expression: $$K_w = a_{H_3O^+}\, a_{OH^-} = 1.00 \times 10^{-14} \text{ at } 25°C$$ where $K_w$ is the autoprotolysis constant for water. This equation does not just apply to pure water, but relates the activities of hydronium and hydroxide ions in all aqueous solutions. |
| **The pH scale** | The pH scale is used to define the acidity of an aqueous solution, where $pH = -\log_{10} a_{H_3O^+}$. When $pH = 7$, the solution is neutral. When $pH > 7$ the solution has an excess of hydroxide ions and is basic, whereas when $pH < 7$ the solution has an excess of hydronium ions and is acidic. An increase in pH corresponds to a decrease in the activity (and concentration) of hydronium ions and an increase in the activity (and concentration) of hydroxide ions. |
| **Related topics** | Fundamentals of equilibria (C1)  Thermodynamics of ions in solution (E2)  Ions in aqueous solution (E1) |

**Brønsted-Lowry theory**

In **Brønsted-Lowry theory**, an **acid** is a **proton ($H^+$) donor** and a **base** is a **proton ($H^+$) acceptor**. Examples of acids are $HCl$, $CH_3COOH$, $H_3O^+$ and $H_2O$. Examples of bases are $NH_3$, $CH_3COO^-$, $H_2O$ and $OH^-$. $H_2O$ can therefore act as either an acid or a base. A reaction between an acid and a base (an **acid–base reaction**) involves the exchange of a proton, for example:

$$HCl + NH_3 \rightarrow Cl^- + NH_4^+$$
$$\text{acid} \quad \text{base}$$

These definitions apply under all conditions, but the most important acid–base systems use water as a solvent. In this case, equilibria are set up in water, which for HA (a general acid) is:

$$HA_{(aq)} + H_2O_{(l)} \rightleftharpoons A^-_{(aq)} + H_3O^+_{(aq)}$$

and for B (a general base) is:

$$B_{(aq)} + H_2O_{(l)} \rightleftharpoons BH^+_{(aq)} + OH^-_{(aq)}$$

In the first equilibrium, the $H_2O$ molecule acts as a base, accepts a proton and forms the **hydronium ion**, $H_3O^+$, which is the hydrated form of the proton in solution. In the second, the $H_2O$ molecule acts as an acid and forms the **hydroxide ion**, $OH^-$. The base that results from the transfer of a proton from the acid is called the **conjugate base** of the acid. Therefore, $A^-$ is the conjugate base of HA and $OH^-$ is the conjugate base of $H_2O$. Similarly, the acid that results from the acceptance of a proton by the base is called the **conjugate acid** of the base. This means that $BH^+$ is the conjugate acid of B and $H_3O^+$ is the conjugate acid of $H_2O$.

**The autoprotolysis constant**

Since water can act as both acid and base, pure water itself ionizes into hydronium and hydroxide ions in the **autoprotolysis equilibrium**,

$$2H_{(2)}O_{(l)} \rightleftharpoons H_3O^+{}_{(aq)} + OH^-{}_{(aq)}.$$

The hydronium ion, $H_3O^+$, is often (inaccurately) represented as a proton, $H^+$, but this is merely equivalent to removing $H_2O$ from both sides of the equation, giving:

$$H_2O_{(l)} \rightleftharpoons H^+{}_{(aq)} + OH^-{}_{(aq)}$$

and so $H^+{}_{(aq)}$ and $H_3O^+{}_{(aq)}$ should be considered to be equivalent.

This is a dynamic equilibrium, which means that protons are continually exchanged between neighboring water molecules. The equilibrium constant (**the autoprotolysis constant** or **water dissociation constant**) is given by:

$$K_w = a_{H_3O^+} \, a_{OH^-}$$

since water is a pure liquid and has an activity of unity (see Topic C1). In fact, this equation does not just apply to pure water, but relates the activities of the hydronium and hydroxide ions in all aqueous solutions. The value of this equilibrium constant is $1.00 \times 10^{-14}$ at 298 K (25°C). This means that for pure water, the concentration of each ion is sufficiently small that their activity coefficients can be approximated to 1 (see Topic C1) and $K_w = c_{H_3O^+} c_{OH^-}/(c^\ominus)^2$.

From this equation it can be calculated that in pure water the concentrations of $H_3O^+$ and $OH^-$ are both equal and are $1.00 \times 10^{-7}$ mol dm$^{-3}$ at 298 K.

**The pH scale**

The equation:

$$pK_w = 14.00 = pH + pOH$$

is obtained by taking the negative logarithm to the base 10 of the autoprotolysis equation, where

$$pK_w = -\log_{10} K_w, \qquad pH = -\log_{10} a_{H_3O^+} \qquad \text{and} \qquad pOH = -\log_{10} a_{OH^-}$$

In general pX denotes $-\log_{10}$ of any given variable, X, so that an increase in X results in a decrease in pX. This equation applies in all aqueous solutions, and it allows the definition of the **pH scale** for water acidity. From the pH of a solution, pOH can always be found from this equation, and the balance of these terms determines the water acidity. For example, when pH = 7, then pOH = 7 and the water solution is **neutral**, as the activities of both $H_3O^+$ and of $OH^-$ are equal at $10^{-7}$. When pH < 7 and pOH > 7, the solution has an excess of $H_3O^+$ and is described as an **acidic** solution. When pH > 7 and pOH < 7, the solution has an excess of $OH^-$ and is described as a **basic** solution. It must be remembered that an increase of one unit in pH corresponds to a ten-fold decrease in the hydronium ion activity and a ten-fold increase in the hydroxide ion activity.

In terms of concentration (see Topics C1, E1 and E2):

$$pH = -\log_{10}\left(\frac{c_{H_3O^+}}{c^\ominus}\right) - \log_{10}\gamma_{H_3O^+}$$

$$and\ pOH = -\log_{10}\left(\frac{c_{OH^-}}{c^\ominus}\right) - \log_{10}\gamma_{OH^-}$$

where $\gamma_i$ is the **activity coefficient** of species $i$.

Often, at low ionic strength (see Topic E2), the $\log_{10}\gamma$ terms are sufficiently small to enable a reasonable estimate of the concentration of the hydronium and hydroxide ions from the pH value:

$$pH \approx -\log_{10}\left(\frac{c_{H_3O^+}}{c^\ominus}\right)$$

$$and\ pOH = 14.00 - pH \approx -\log_{10}\left(\frac{c_{OH^-}}{c^\ominus}\right).$$

# C3 FURTHER ACIDS AND BASES

## Key Notes

**Strong and weak acids and bases**

Strong acids completely dissociate into their conjugate base and hydronium ions. Strong bases completely dissociate into their conjugate acid and hydroxide ions. For weak acids and bases, incomplete dissociation occurs and an acid–base equilibrium is established.

**The acidity constant**

The equilibrium constant for the dissociation of a weak acid is called the acid dissociation constant or acidity constant, $K_a$. The equilibrium constant for the dissociation of a weak base is called the base dissociation constant or basicity constant, $K_b$. The acid dissociation constant of an acid is linked to the base dissociation constant of its conjugate base by the equation $K_a K_b = K_w$. $pK_a$ and $pK_b$ are a measure of the acid and base strength with lower values meaning increased strength. $pK_a$ and $pK_b$ correspond to the pH and pOH values respectively when the acid and conjugate base activities are equal. A strong acid results in a very weak conjugate base and a strong base results in a very weak conjugate acid.

**Salt solutions**

Acid dissociation constants can be used to predict whether a salt solution will be acidic or basic. Salts often consist of an acid (the positive ion) and a base (the negative ion). If the $pK_a$ of the acid is less than the $pK_b$ of the base, then the solution will be acidic. If the converse is true, the solution will be basic.

**Polyprotic acids and bases**

A polyprotic acid is an acid that can donate more than one proton. Acid dissociation can generally be considered to be a number of stepwise single proton dissociation reactions, each with their own acid dissociation constant. Combining these stepwise constants produces the multiproton acid dissociation constant. Multiproton base dissociation reactions (for a polyprotic base) can also be considered to be a succession of single proton base dissociation reactions.

**Related topics**

Fundamentals of equilibria (C1)
Ions in aqueous solution (E1)

Electrochemistry and ion concentration (E5)

**Strong and weak acids and bases**

A **strong acid** is an acid for which complete dissociation can be assumed to occur, forming the hydronium ion and the conjugate base. A good example is:

$$HCl_{(aq)} + H_2O_{(l)} \rightarrow H_3O^+_{(aq)} + Cl^-_{(aq)}$$

**Weak acids** show much less tendency to form hydronium ions and complete dissociation does not occur. Instead, an acid dissociation equilibrium is established and there is significant undissociated acid in solution. An example is the ammonium ion:

$$NH_4^+_{(aq)} + H_2O_{(l)} \rightleftharpoons NH_{3(aq)} + H_3O^+_{(aq)}$$

These equilibria have equilibrium constants called **acidity constants** or **acid dissociation constants**. It therefore follows that a **strong base** is a base that completely dissociates into hydroxide ion and its conjugate acid. An example is potassium hydroxide:

$$KOH_{(aq)} \rightarrow K^+_{(aq)} + OH^-_{(aq)}$$

A **weak base** does not dissociate completely and, as with a weak acid, establishes an equilibrium. An example is ammonia, which is the conjugate base of the weak acid given above:

$$NH_{3(aq)} + H_2O_{(l)} \rightleftharpoons NH^+_{4(aq)} + OH^-_{(aq)}$$

**The acidity constant**

The general proton transfer equilibrium between an acid, HA, and water:

$$HA_{(aq)} + H_2O_{(l)} \rightleftharpoons A^-_{(aq)} + H_3O^+_{(aq)}$$

has an equilibrium constant called the **acidity constant** or **acid dissociation constant**, $K_a$, given by (see Topic C1):

$$K_a = \frac{a_{H_3O^+} a_{A^-}}{a_{HA}}$$

The smaller the value of $pK_a$ (or as $pK_a = -\log_{10}K_a$, the larger the value of $K_a$) the further the equilibrium position is towards the right-hand side and the more the acid is dissociated. Thus $pK_a$ is a measure of the **acid strength**, or the ability of the acid to donate protons and the stronger the acid, the smaller is its value of $pK_a$.

By taking logarithms of both sides, this equation rearranges to:

$$pH = pK_a + \log_{10}\left(\frac{a_{A^-}}{a_{HA}}\right)$$

which is the **Henderson-Hasselbach equation** (see Topic C4).

From this equation, $pK_a$ is the pH at which the activity of the acid, HA, and its conjugate base, $A^-$, are equal, when the logarithmic term becomes equal to zero.

**Base dissociation constants** or **basicity constants**, $K_b$, can also be used for base equilibria. The base dissociation constant is the equilibrium constant for the general reaction:

$$B_{(aq)} + H_2O_{(l)} \rightleftharpoons BH^+_{(aq)} + OH^-_{(aq)}$$

which is equivalent to

$$A^-_{(aq)} + H_2O_{(l)} \rightleftharpoons AH_{(aq)} + OH^-_{(aq)}$$

as the conjugate base of an acid, $A^-$, is a base, B, and so $B \equiv A^-$. This means that since the conjugate acid of a base, B, is an acid and $BH^+ \equiv HA$ these symbols can be used interchangeably. The base dissociation constant is given by the equation (see Topic C1):

$$K_b = \frac{a_{BH^+} a_{OH^-}}{a_B}$$

By analogy with $K_a$, the greater the magnitude of $K_b$ (and the smaller the size of $pK_b$), the greater is the **base strength**, or the ability of the base to accept a proton. The base dissociation constant of any base ($B \equiv A^-$) can simply be calculated from $K_a$ for its conjugate acid (HA ≡ BH$^+$) using the relationship $K_aK_b = K_w$ or p$K_a$ + p$K_b$ = p$K_w$ = 14.00.

This means that tabulation of both p$K_a$ values for acids and p$K_b$ values for their conjugate bases is unnecessary and often only p$K_a$ values for a range of acids are given. Also from this relationship, it is clear that as the strength of an acid is increased, the base strength of its conjugate base is decreased, and as the strength of a base is increased, the acid strength of its conjugate acid is decreased. This means that as HCl is a strong acid, Cl$^-$ is a very weak base.

By analogy, the relationship pOH = p$K_b$ + $\log_{10}$ ($a_B/a_{BH^+}$) can be derived from the above equation for the base dissociation equilibrium, which means that pOH = p$K_b$ when the activities of the base and its conjugate acid are equal.

**Salt solutions**

Acidity constants provide an easy means of predicting whether a solution formed from dissolving a salt will be acidic or basic. These solutions generally consist of an acid (the positive ion or cation, see Topic E1) and a base (the negative ion or anion). The pH of the resulting solution will be determined by the relative strengths of the acid and base, with the strongest dominating. For example, ammonium hydroxide consists of a relatively weak acid (NH$_4^+$) and a strong base (OH$^-$) and on dissociation:

$$NH_4OH_{(aq)} \rightarrow NH_{4(aq)}^+ + OH^-_{(aq)}$$

the solution will be basic, as the proton accepting ability of the hydroxide ion (as measured by its very small p$K_b$ value) will dominate the proton donating ability of the ammonium ion (as measured by its relatively large p$K_a$ value). This means that overall, more hydroxide ions will be present than hydronium ions in solution. In contrast, a solution of ammonium chloride will be acidic, because on dissociation:

$$NH_4Cl_{(aq)} \rightarrow NH_{4(aq)}^+ + Cl^-_{(aq)}$$

the ammonium ion has greater strength as an acid than the chloride ion has as a base and more hydronium ions will be present in solution.

**Polyprotic acids and bases**

A **polyprotic acid** is an acid that can donate more than one proton. An example is carbonic acid, $H_2CO_3$. A series of stepwise acid dissociation equilibria are set up in these systems, each of which involves donation of a single proton. For carbonic acid, these are:

$$H_2CO_{3(aq)} + H_2O_{(l)} \rightleftharpoons HCO^-_{3(aq)} + H_3O^+_{(aq)} \qquad K_{a1} = \frac{a_{HCO_3^-} a_{H_3O^+}}{a_{H_2CO_3}}$$

$$HCO^-_{3(aq)} + H_2O_{(l)} \rightleftharpoons CO^{2-}_{3(aq)} + H_3O^+_{(aq)} \qquad K_{a2} = \frac{a_{CO_3^{2-}} a_{H_3O^+}}{a_{HCO_3^-}}$$

Generally, the stepwise acid dissociation constants are denoted $K_{ax}$ as shown, where the $x$th proton is removed from the polyprotic acid in each equilibrium. $K_{ax} > K_{a(x+1)}$ and hence p$K_{ax} <$ p$K_{a(x+1)}$, which means the acid strength decreases for each successive deprotonation. This is due to the increased negative charge on the acid after each deprotonation, which electrostatically attracts the remaining proton(s) and decreases its (their) tendency to be donated.

At 25°C for carbonic acid, p$K_{a1}$ = 6.37 and p$K_{a2}$ = 10.25, whilst for phosphoric acid, p$K_{a1}$ = 2.12, p$K_{a2}$ = 7.21 and p$K_{a3}$ = 12.67. Generally, p$K_{ax}$ values are sufficiently widely separated (by greater than 2 units) that each proton transfer can be considered to occur independently and sequentially.

The relationship $pH = pK_{ax} + \log_{10}(a_{A^-}/a_{HA})$ can therefore be applied to each stepwise dissociation, with $a_{A^-}$ and $a_{HA}$ being the activities of the conjugate base and acid in the the $x$th dissociation equilibrium. A consequence is that when the pH of the solution is at or near $pK_{ax}$, often it can be assumed that only the $x$th single proton dissociation equilibrium is occurring.

The acid dissociation constant for the overall multiproton donation equilibrium, if required, is simply given by the product of the stepwise acid dissociation constants, for example:

$$H_2CO_{3(aq)} + 2H_2O \rightleftharpoons CO_{3(aq)}^{2-} + 2H_3O^+_{(aq)} \qquad K_a = \frac{a_{CO_3^{2-}}\, a^2_{H_3O^+}}{a_{H_2CO_3}}$$

$$K_a = K_{a1}K_{a2} \text{ or more generally, } K_a = \prod_x K_{ax}$$

The same arguments apply to polyprotic bases, which can be treated in a similar manner. For example, for:

$$CO_{3(aq)}^{2-} + 2H_2O \rightleftharpoons H_2CO_{3(aq)} + 2OH^-_{(aq)} \qquad K_b = \frac{a_{H_2CO_3}\, a^2_{OH^-}}{a_{CO_3-}}$$

which is the multiproton conjugate base dissociation reaction for carbonic acid, the single proton reactions are:

$$CO_{3(aq)}^{2-} + H_2O \rightleftharpoons HCO_{3(aq)}^- + OH^-_{(aq)} \qquad K_{b1} = \frac{a_{HCO_3^-}\, a_{OH^-}}{a_{CO_3^{2-}}}$$

and

$$HCO_{3(aq)}^- + H_2O \rightleftharpoons H_2CO_{3(aq)} + OH^-_{(aq)} \qquad K_{b2} = \frac{a_{H_2CO_3}\, a_{OH^-}}{a_{HCO_3^-}}$$

As with polyprotic acids, $K_b = K_{b1}K_{b2}$ or, more generally, $K_b = \prod_x K_{bx}$, and the $K_{bx}$ values are usually sufficiently widely separated that they can be considered as a series of distinct single proton reactions, with:

$$pOH = pK_{bx} + \log_{10}\left(\frac{a_{BH^+}}{a_B}\right)$$

where $a_{BH^+}$ and $a_B$ are the activities of the conjugate acid and base respectively in the $x$th base dissociation equilibrium. This also means that $K_{ax}K_{bx} = K_w$ for the $x$th single proton dissociation equilibrium.

# C4 ACID–BASE TITRATIONS

## Key Notes

**Titration curves**

An acid–base titration involves the addition of a titrant solution to an analyte solution. One of these solutions contains an acid, whilst the other contains a base; the titrant is usually a strong acid or strong base. A pH titration curve of pH versus volume of titrant is produced. The end point of the reaction is when the reaction between the acid and the base has gone to completion and each has neutralized the other, which corresponds to a sharp change in pH. The end point is at pH = 7 for a strong acid–strong base titration, at pH > 7 for a weak acid–strong base titration and at pH < 7 for a weak base–strong acid titration.

**Buffers**

A buffer solution contains large and equal concentrations of an acid and its conjugate base. The pH of this solution is approximately equal to the $pK_a$ of the acid. Addition of small amounts of acid or base results in the mopping up or the release of protons by the conjugate base or the acid as necessary, which keeps the solution pH constant.

**Acid–base indicators**

An acid–base indicator is a molecule that is differently colored in its acid and its conjugate base forms. The indicator changes color when the pH of the solution changes between $pK_a - 1$ (the acid form) and $pK_a + 1$ (the base form). The color change of an indicator can be used to determine the end point of an acid–base titration providing its $pK_a$ coincides with the pH at the end point.

**Related topics**    Fundamentals of equilibria (C1)    Further acids and bases (C3)

**Titration curves**

An **acid–base titration** is an extremely useful experimental method for deter-mining the $pK_a$ and also the amount of an unknown acid or base in solution. In an acid–base titration, the pH of the solution is measured, generally with a glass electrode (see Topic E5), as reaction between an acid and a base occurs. This is achieved by using an acid solution and a base solution and adding one to the other. The solution that is added is called the **titrant solution**; this is added to the **analyte solution**, and the whole process is called **titration**. Generally the titrant is a strong acid or a strong base as appropriate. The results are plotted as the pH of the solution versus the volume of titrant added, called a **pH titration curve**.

When the **stoichiometry** (or ratio of number of moles) of acid and base in the solution is such that exact **neutralization** occurs (acid and base have reacted to produce a salt solution), for example:

$$H_2CO_{3(aq)} + 2NaOH_{(aq)} \rightarrow Na_2CO_{3(aq)} + 2H_2O_{(aq)}$$

this is called the **stoichiometric point**, or the **equivalence point**, or the **end point** of the titration. If the stoichiometry of the reaction is unknown, this can then be calculated from the volumes and concentrations of the titrant and analyte solutions at the end point (as this gives the number of moles of analyte and titrant, see Topic D1).

For a strong acid–strong base titration, for example:

$$H_3O^+_{(aq)} + Cl^-_{(aq)} + Na^+_{(aq)} + OH^-_{(aq)} \rightarrow Na^+_{(aq)} + Cl^-_{(aq)} + 2H_2O_{(l)}$$
strong acid (HCl)     strong base (NaOH)

both the strong acid and the strong base are completely dissociated and so the reaction is simply $H_3O^+_{(aq)} + OH^-_{(aq)} \rightarrow 2H_2O_{(l)}$, with the other ions, $Na^+$ and $Cl^-$, taking no part in the reaction. At the end point, the solution is neutral (pH = 7). If the acid is the analyte and the base is the titrant, the pH curve starts at a low pH and rises sharply through the end point of pH = 7 (*Fig. 1a*). However, if the base is the analyte and the acid is the titrant, the pH starts at a high pH and falls sharply through the end point of pH = 7 (*Fig. 1c*).

For strong acid–strong base titrations, as the change in pH with volume of added titrant, $V$, is steepest at the end point, the end point can often be most accurately located by plotting $dpH/dV$, the change in pH with respect to $V$, against $V$. The end point then corresponds to a sharp maximum (for rising pH with $V$, when titrating acid analyte with base, *Fig. 1b*) or minimum (for falling pH with $V$ when titrating base analyte with acid, *Fig. 1d*) in the plot, which allows precise end point location.

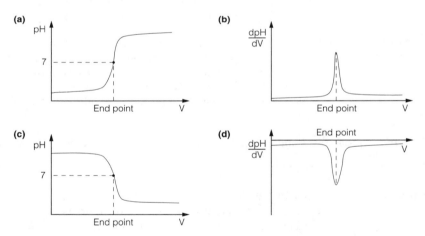

*Fig. 1.    The titration reactions of a strong base and a strong acid. (a) pH titration curve for reaction of an acid analyte with a base titrant; (b) corresponding dpH/dV plot; (c) pH titration curve for reaction of a base analyte with an acid titrant; (d) corresponding dpH/dv plot.*

For the titration of a weak acid (analyte) by a strong base (titrant), for example:

$$H_2CO_{3(aq)} + 2Na^+_{(aq)} + 2OH^-_{(aq)} \rightarrow 2Na^+_{(aq)} + CO_3^{2-}_{(aq)} + 2H_2O_{(l)}$$
weak acid     strong base (2 NaOH)

at the end point the salt solution is weakly basic (pH > 7), as the conjugate base of a weak acid is stronger than the conjugate acid of a strong base (*Fig. 2a*; see also Topic C3). Furthermore, when strong base is added before the end point, the combined equilibrium between $H_2CO_3$ and $CO_3^{2-}$ in the weak acid solution:

$$H_2CO_{3(aq)} + 2H_2O \rightleftharpoons CO_3^{2-}_{(aq)} + 2H_3O^+_{(aq)}$$

shifts to the right towards the production of more hydronium ions, according to **Le Chatelier's principle** (see Topic C1), offsetting the rise in OH⁻ concentration and slowing the change in pH. Therefore the steepest rise in pH occurs at the end point, where there is no acid left to offset the rise in OH⁻ concentration. This means that a graph of dpH/dV against V can again be used to determine the end point, which is where a maximum in dpH/dV occurs.

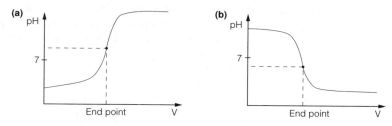

Fig. 2.    pH titration curves for the reaction of (a) a weak acid solution (as analyte) with a strong base solution (as titrant), (b) a weak base solution (as analyte) with a strong acid solution (as titrant).

For the titration of a weak base (analyte) by a strong acid (titrant), for example:

$$NH_{3(aq)} + H_3O^+_{(aq)} + Cl^-_{(aq)} \rightarrow NH_4^+_{(aq)} + Cl^-_{(aq)} + H_2O_{(l)}$$
$$\text{strong acid (HCl)}$$

at the end point the salt solution is weakly acidic (pH < 7), as the conjugate acid of a weak base is stronger than the conjugate base of a strong acid (*Fig. 2b*; see also Topic C3).

Again, before the end point, the fall in pH due to the addition of the strong acid is partially offset by a shift in the weak base equilibrium:

$$NH_{3(aq)} + H_2O_{(l)} \rightleftharpoons NH_4^+_{(aq)} + OH^-_{(aq)}$$

to the right according to Le Chatelier's principle (see Topic C1), so that the change in pH with added acid is greatest at the end point.

Again, a graph of dpH/dV against V could be used to determine the end-point, as this is where a minimum in dpH/dV should occur. An alternative method for determining these end points does not involve the measurement of pH, but instead involves the use of an **acid–base indicator**.

**Buffers**

The ability of a weak acid to partially offset the rise in pH caused by the addition of a base and of a weak base to partially offset the fall in pH due to the addition of an acid is exploited in **buffer solutions**. These consist of a solution containing large and equal concentrations of both a weak acid, HA, and its conjugate weak base, A⁻, for which (see Topic C3) pH = p$K_a$ + log₁₀ (a$_{A^-}$/a$_{HA}$), which in terms of concentrations is (see Topic C1):

$$pH = pK_a + \log_{10}\left(\frac{c_{A^-}}{c_{HA}}\right) + \log_{10}\left(\frac{\gamma_{A^-}}{\gamma_{HA}}\right)$$

where $c_{HA}$ and $c_{A^-}$ are the concentrations of acid and conjugate base, respectively, and $\gamma_{HA}$ and $\gamma_{A^-}$ are their respective activity coefficients. The concentrations of acid and base are typically of the order of 0.1 M such that the activity coefficients cannot be approximated to unity, but the effect of the activity term is generally small and hence:

$$pH \approx pK_a + \log_{10}\left(\frac{c_{A^-}}{c_{HA}}\right)$$

This is known as the **Henderson-Hasselbalch equation**. For a buffer solution, $c_{A^-} = c_{HA}$, and so $pH \approx pK_a$. As large concentrations of HA and $A^-$ are present, addition (or production) of a relatively small amount of base compared with the amount of HA present results in the reaction $HA_{(aq)} + OH^-_{(aq)} \rightarrow A^-_{(aq)} + H_2O_{(l)}$. This mops up the added hydroxide ion, whilst causing little change to the large values of $c_{HA}$ or $c_{A^-}$, and hence little change to the solution pH. Similarly, addition (or production) of a relatively small amount of acid produces the reaction $A^-_{(aq)} + H_3O^+_{(aq)} \rightarrow HA_{(aq)} + H_2O_{(l)}$ which mops up the added acid, whilst maintaining $c_{HA}$, $c_{A^-}$ and hence the solution pH constant. Judicious choice of the acid/conjugate base pair therefore allows the pH of a solution to be maintained at a desired value, determined by the $pK_a$ of the acid, even if relatively small amounts of hydroxide or hydronium ions are being added to or removed from the solution (*Fig. 3*). This is termed a **buffered solution**.

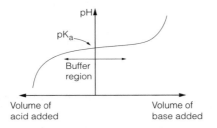

*Fig. 3. pH response of a buffered solution to the addition of acid or base.*

Most biological systems are buffered solutions, with their pH maintained at or around a value of 7, the optimum value for physiological processes, despite the presence of variable amounts of acid–base species such as dissolved carbon dioxide (carbonic acid).

**Acid–base indicators**

An **acid–base indicator** is generally a large, soluble organic molecule which in its acid form (HIn) is colored and in its conjugate base form (In$^-$) is differently colored. The **Henderson-Hasselbalch equation** for this species is:

$$pH \approx pK_a + \log_{10}\left(\frac{c_{In^-}}{c_{HIn}}\right)$$

and so if the solution pH changes from a value much less than $pK_a$ where $c_{HIn} \gg c_{In^-}$ to one much greater than $pK_a$ where $c_{HIn} \ll c_{In^-}$, the indicator changes from its acid form (HIn) to its basic form (In$^-$) and changes color. In fact this change is generally seen to take place between $pH = pK_a - 1$, where there is a ten-fold excess of HIn over In$^-$ and $pH = pK_a + 1$, where there is a ten-fold excess of In$^-$ over HIn. The abrupt change in pH at the end point of an acid–base titration is at least as large as two pH units and so the color change of a small amount of indicator added to the acid–base titration can be used to detect this end point. This will be possible as long as the pH at the end point is approximately equal to the $pK_a$ of the indicator.

It is important that the concentration of indicator is very much smaller than the concentration of acid and of base in the titration. This ensures that very little extra titrant is required to effect the indicator acid–base color change, which ensures the accuracy of the end-point determination is unaffected. This can easily be achieved, as indicators are highly colored.

# C5 SOLUBILITY

---

## Key Notes

**Solubility equilibria**

Partially soluble salts only partly dissolve in solution. An equilibrium between the ions and the solid salt is established and a saturated solution of the ions is produced. The equilibrium dissociation constant for this process is called the solubility product, $K_{sp}$. For partially soluble salts, the solubility of the salt, $s$, is simply determined by $K_{sp}$.

**The common ion effect**

When a common ion (an ion which is part of the equilibrium reaction) is added to the solution, the solubility of the salt decreases. This is consistent with Le Chatelier's principle, as the equilibrium position changes to remove the ion from solution.

**The inert ion effect**

When an inert ion (which takes no part in the equilibrium reaction) is added, the solubility of the salt increases. This is due to the energetically favorable electrostatic interactions between the inert ions and the salt ions, which stabilize the ions in solution, favoring the dissociation of more salt.

**Related topics**

Fundamentals of equilibria (C1)
Ions in aqueous solution (E1)

Thermodynamics of ions in
solution (E2)

---

**Solubility equilibria**

**Partially** (or **sparingly**) **soluble salts** are salts that only partly dissolve, forming a **saturated solution** of ions. For these systems an equilibrium exists between the solid salt and the dissolved ions:

$$MX_{(s)} \rightleftharpoons M^+_{(aq)} + X^-_{(aq)}$$

The equilibrium constant for this reaction is often called the **solubility product, $K_{sp}$**, and is given by:

$$K_{sp} = a_{M^+} a_{X^-}$$

since MX is a pure solid (see Topic C1), where $a_{M^+}$ and $a_{X^-}$ are the activities of the $M^+$ ion (the **cation**, see Topic E1) and the $X^-$ ion (the **anion**, see Topic E1). A good example would be solid silver chloride, which only partially dissolves into $Ag^+$ and $Cl^-$ ions. By substituting for the activity of the ions (see Topics C1 and E2):

$$K_{sp} = \frac{c_{M^+} c_{X^-}}{\left(c^{\ominus}\right)^2} \left(\gamma_{M^+} \gamma_{X^-}\right)$$

where $\gamma_{M^+}$ and $\gamma_{X^-}$ are the activity coefficients of $M^+$ and $X^-$ and $c^{\ominus}$ is the standard concentration of 1 mol dm$^{-3}$. For **sparingly soluble salts**, such as silver chloride, which have concentrations much less than 0.001 mol dm$^{-3}$, there is negligible interaction between ions in solution and the activity coefficients can be approximated to unity. The equation then becomes:

$$K_{sp} = \frac{c_{M^+} c_{X^-}}{\left(c^{\ominus}\right)^2}$$

The **solubility, s**, of the salt is the concentration of dissolved salt in the solution. For a salt MX, $s = c_{M^+} = c_{X^-}$ as one mole of $M^+$ and $X^-$ ions is produced by the dissolution of one mole of salt. Therefore:

$$K_{sp} = \frac{s^2}{\left(c^\ominus\right)^2} \text{ and } s = \sqrt{K_{sp}}\, c^\ominus \text{ or } s = \sqrt{K_{sp}} \text{ mol dm}^{-3}$$

which allows the solubility of the salt in water to be determined from $K_{sp}$. For sparingly soluble salts containing ions with differing stoichiometries, a similar expression can be obtained. For example for silver sulfide, $Ag_2S$, the solubility equilibrium is:

$$Ag_2S_{(s)} \rightleftharpoons 2Ag^+_{(aq)} + S^{2-}_{(aq)} \text{ and } K_{sp} = \frac{c^2_{Ag^+} c_{S^{2-}}}{\left(c^\ominus\right)^3}$$

One mole of silver sulfide dissolves to form one mole of sulfide ions and two moles of silver ions.

Therefore $s = c_{S^{2-}} = \tfrac{1}{2}c_{Ag^+}$, $K_{sp} = \dfrac{4s^3}{\left(c^\ominus\right)^3}$ and $s = \left(K_{sp}/4\right)^{\frac{1}{3}} c^\ominus$ or $s = \left(K_{sp}/4\right)^{\frac{1}{3}}$ mol

**The common ion effect**

The **common ion effect** considers the effect on the solubility of the salt MX of adding either $M^+$ or $X^-$. An example is the addition of NaCl to a saturated AgCl solution. **Le Chatelier's principle** predicts that the equilibrium:

$$AgCl_{(s)} \rightleftharpoons Ag^+_{(aq)} + Cl^-_{(aq)}$$

will shift to the left to counteract the increase in chloride ion concentration and that the solubility will decrease. Quantitatively, if a concentration, $c$, of NaCl is added which is enough to swamp the original concentration of $Cl^-$ in solution, then $c_{Cl^-} = c$. The solubility of the salt would then be given by $s = c_{Ag^+}$, as only the silver ions in solution must have come from the silver chloride salt. Hence:

$$K_{sp} = \frac{c_{Ag^+} c_{Cl^-}}{\left(c^\ominus\right)^2} = \frac{sc}{\left(c^\ominus\right)^2}$$

and

$$s = \frac{K_{sp}\left(c^\ominus\right)^2}{c}$$

This confirms the shift to the left of the equilibrium with the solubility, $s$, decreasing as $c$ increases. This is the **common ion effect**.

It must be remembered that this equation only rigorously applies if c is sufficiently small (of the order of 0.001 mol dm$^{-3}$ or less) to ensure that there is no interaction between the ions in solution. If $c$ becomes larger than this, the energetically favorable interactions between ions seen in the **inert ion effect** become increasingly important (see Topic E1) and the effects of activity cannot be neglected. In this case, the solubility equation becomes

$$s = \frac{K_{sp}\left(c^\ominus\right)^2}{c}\left(\frac{1}{\gamma_{Ag^+}\gamma_{Cl^-}}\right)$$

which as $\gamma < 1$ (see Topic E2) allows for the small increase in solubility due to the electrostatic stabilization of the ions. However, this effect is relatively small and is dominated by the common ion effect when adding a common ion and so even at higher concentrations, an overall decrease in solubility is seen in this case.

**The inert ion effect**

When **inert ions**, which take no part in the solubility equilibrium, are added to the solution, these tend to cause an increase in the solubility of the salt, called the **inert ion effect**. An example is adding an $NaNO_3$ solution to the saturated AgCl solution. In this case

$$K_{sp} = \frac{c_{Ag^+} c_{Cl^-}}{(c^{\ominus})^2} \left( \gamma_{Ag^+} \gamma_{Cl^-} \right)$$

and since

$$s = c_{Ag^+} = c_{Cl^-},$$

$$s = \sqrt{K_{sp}} \, c^{\ominus} \left( \frac{1}{\gamma_{Ag^+} \gamma_{Cl^-}} \right)^{\frac{1}{2}}$$

As the concentration of inert ions is increased to 0.001 M and above, the effects of electrostatic ion interaction become increasingly important, stabilizing the ions in solution (see Topics E1 and E2) and leading to greater ion dissociation. Thus the activity coefficients, $\gamma$, become significantly less than unity and the solubility, $s$, increases. Values of $\gamma$ at any ion concentration can be calculated by using **Debye-Hückel theory** (see Topic E2), which allows the calculation of $s$.

# D1 NON-ELECTROLYTE SOLUTIONS

## Key Notes

**Composition**

A solution is a mixture of one or more solute(s) (the minority species) dispersed in a solvent (the majority species). Usually, the solvent, and hence the solution, is a liquid. In non-electrolyte solutions, the species are not charged and cannot interact electrostatically. The composition, which is the relative amount of solute(s) and solvent in the system, can be defined by the concentration, $c_i$, of each species, $i$. Alternative but related measures of composition are the mole fraction, $x_i$, and the molality, $m_i$, of a species. Mole fraction is used more generally for all mixtures, even those where a solution is not formed. Molality is only rarely used.

**Chemical potential**

The chemical potential, $\mu_i$, is the partial molar Gibbs free energy of a species, $i$. The total Gibbs free energy of any mixture is obtained by combining the chemical potentials of all the constituent species. As with other partial molar properties, the chemical potential of pure $i$ is usually not equal to the chemical potential of $i$ in a mixture, due to differences in the molecular environment. These differences in chemical potential are given by the variation in the activity of $i$, $a_i$, which is related to the chemical potential by the equation $\mu_i = \mu^{\ominus}_i + RT\ln a_i$, where $\mu^{\ominus}_i$ is the standard chemical potential at an activity of unity.

**Related topics**

Perfect gases (A1)  
Free energy (B6)  
Solutions (D2)

Phase equilibria (D4)  
Phase diagrams of mixtures (D5)

**Composition**

A **solution** is a mixture of two or more species; this consists of one or more minority substances, the **solute(s)**, dispersed in a majority substance, present in greater amounts, the **solvent**. In the vast majority of solutions, the solvent, and hence the solution, is a liquid (although solid solutions are possible). The term **mixture** can also be used more generally to describe a system with more than one substance, often under conditions that include approximately equal amounts, where no one substance can be considered the solvent (see Topics D4 and D5). It is easiest to consider chemical species that have no charge (are not **ions**) and cannot interact **electrostatically**, termed **non-electrolyte solutions**. Electrostatic interaction is the added complication of the relatively long-range attractive and repulsive forces between ions of opposite and like charges found in **ionic** or **electrolyte solutions**; these are considered in Section E.

**Concentration** is the normal variable used to define the **composition**, or the relative amounts of solvent and solute in a solution. The concentration of a species A, $c_A$, is defined as:

$$c_A = \frac{n_A}{V}$$

where $n_A$ is the number of moles of A in solution and $V$ is the volume of the solution. This normally has units of mol dm$^{-3}$ (also written M for **molar**). The concentration of A is often also written as [A]. A solution is typically prepared by dispersing (**dissolving**) $n_A$ moles of solute in solvent to produce a final total volume of $V$. This is generally not the same as mixing $n_A$ moles of solute with a volume, $V$, of water, as the volume after mixing is usually not $V$ in this case.

An alternative measure of the composition of a mixture is the **mole fraction,** $x_A$. The mole fraction is given by:

$$x_A = \frac{n_A}{n}$$

where $n$ is the total number of moles of species present in the mixture. For example, for a solution containing two solutes, A and B, and a solvent, C:

$$n = n_A + n_B + n_C$$

where $n_i$ is the number of moles of a species, $i$. By dividing both sides by $n$, this means that $x_A + x_B + x_C = 1$ or more generally, $\Sigma_i x_i = 1$.

Concentration and mole fraction are clearly closely related, as $V$ is related to $n$. Concentration is the more frequently used composition variable, but mole fraction is more general and is often preferred to concentration in mixtures where there is no obvious solvent. These systems are commonly found in phase diagrams, which is why mole fraction is the composition variable of choice in the phase diagrams of mixtures.

A third composition variable used occasionally for solutions is the **molality** of species A, $m_A$, defined by:

$$m_A = \frac{n_A}{M}$$

where $M$ is the total mass of the solute and solvent. When water is used as a solvent, 1 dm$^3$ (1000 cm$^3$) of water has a mass of 1.000 kg at room temperature (298 K). This means that the molality of a solute (in mol kg$^{-1}$) is approximately equal to its concentration (in mol dm$^{-3}$) for dilute aqueous solutions, where the contribution of the mass of the solute to the overall mass is negligible.

**Chemical potential**

Partial molar quantities can be calculated for many variables, including thermodynamic variables. The most important thermodynamic variable for a **closed system**, which can exchange energy with its surroundings, is the **Gibbs free energy** (see Topic B6). The **partial molar Gibbs free energy** or **chemical potential,** $\mu_i$, of a species is the Gibbs free energy per mole of the species in the mixture. Therefore, the total Gibbs free energy, $G$, of a mixture of species of $n_A$ moles of A, $n_B$ moles of B and $n_C$ moles of C is given by $G = n_A \mu_A + n_B \mu_B + n_C \mu_C$ or, more generally, $G = \Sigma_i n_i \mu_i$ for all the species, $i$, in the system. As with all other partial molar quantities, the chemical potential of a pure substance is generally not the same as the chemical potential of that substance in a mixture. This is due to differences in the molecular arrangement, which produce differences in the molecular interactions, in the two systems.

Generally, these differences in chemical potential for any species $i$ at any temperature, $T$, are given by its **activity, $a_i$,** as:

$$\mu_i = \mu_i^{\ominus} + RT\ln a_i$$

where $\mu_i^{\ominus}$ is the standard chemical potential of the species, or the chemical potential when the activity is unity. For a perfect gas (see Topic A1), the activity is given by $a_i = p_i/p^\ominus$ where $p_i$ is the **partial pressure** of the gas, $i$, (see Topic A1) and $p^\ominus$ is the standard pressure of 1 atmosphere. Essentially, the more chemical potential the molecules in a perfect gas have (the more Gibbs free energy) the faster they move and the more pressure they exert. This is a relatively simple expression, which is a consequence of the fact there are no intermolecular interactions in perfect gases. Generally, in all systems, the activity expression allows the change in the chemical potential (the partial molar Gibbs free energy) of any species to be calculated when its molecular environment is changed from standard conditions. However, for more complicated systems which have significant intermolecular interactions, such as ions in solution (see Topics E1 and E2) or non-ideal mixtures of liquids (see Topic D2), the activity relationship is more complicated, reflecting the greater complexity introduced by these interactions.

At equilibrium, the change in Gibbs free energy for the reaction is zero (see Topic C1) and hence the Gibbs free energy of reactants and products are equal. For a physical transition, for example vaporization:

$$A_{(l)} \rightleftharpoons A_{(g)}$$

this means that the chemical potentials of A in the gas phase and A in the liquid phase must be equal or $\mu_{(l)} = \mu_{(g)}$.

Also, away from equilibrium, the overall Gibbs free energy of any reaction

$$aA + bB \rightarrow cC + dD$$

can be calculated from the individual chemical potentials as

$$\Delta G = (c\mu_C + d\mu_D) - (a\mu_A + b\mu_B).$$

# D2 SOLUTIONS

## Key Notes

**Ideal solutions**

An ideal solution is a mixture of two species, A and B, which show similar molecular interactions between molecules of A, molecules of B and molecules of B and A. By definition, an ideal solution obeys Raoult's Law, $p_i = x_i p_i^*$, where $p_i$, $x_i$ and $p_i^*$ are the partial vapor pressure, the mole fraction and the vapor pressure of liquid species $i$, and $i$ is either A or B.

**Non-ideal solutions**

The vast majority of solutions are non-ideal solutions and show deviation from Raoult's law. In this case, for dilute solutions of a solute B in a solvent A, termed an ideal-dilute solution, Raoult's law applies to the solvent. For the solute, the partial vapor pressure, $p_B$, is related to its mole fraction, $x_B$, by Henry's law:

$$p_B = K_B x_B$$

where $K_B$ is the Henry's law constant, which quantifies the deviation from ideal behavior of B. $K_B$ is a constant for a particular solute B in a particular solvent, A.

**Related topics**

Free energy (B6)                    Colligative properties (D3)
Fundamentals of equilibria (C1)

---

**Ideal solutions**

The mole fraction of any species in a liquid system is the equivalent variable to the partial pressure of a species in a gas, as in each case increasing this variable causes an increase in the number of molecules of the species per unit volume. This means that the activity of a liquid is related to its mole fraction. An **ideal solution** of a mixture of two liquids, A and B, is one in which the interactions between similar pairs of molecules, A and A or B and B in a solution are similar in magnitude to those between the dissimilar molecules A and B. A good example is benzene and toluene, which are molecules with very similar sizes and shapes and have very similar interactions. In this case **Raoult's law** is obeyed, which is:

$$p_i = x_i p_i^*$$

where $p_i$, $x_i$ and $p_i^*$ are the **partial vapor pressure**, the mole fraction and the **vapor pressure** of liquid species $i$, where $i$ is A or B. The vapor pressure of species $i$ is the pressure of gas $i$ in **equilibrium** with the pure liquid species, $i$ (see Topic C1). The partial vapor pressure of liquid $i$ is therefore the partial pressure of $i$ in the vapor mixture in equilibrium with the liquid mixture. Generally the vapor pressure, $p$, of the vapor mixture is given by $p = \Sigma p_i$.

The origin of Raoult's law is that the partial vapor pressure of $i$ is due to an equilibrium at the surface between the molecules of $i$ in the liquid vaporizing and the molecules of $i$ in the vapor condensing. This reaction occurs over that fraction of the surface covered by $i$, which is $x_i$, and so $p_i \propto x_i$, and when $x_i = 1$ (for pure $i$) $p_i = p_i^*$.

**Non-ideal solutions**

Raoult's law only applies to a very few systems at all compositions. Generally, it is very rare for the interactions between A and B to be exactly the same or even similar. This greatly complicates the situation for high concentrations of solute. However, for all solutions where the solute (B) is at a very low concentration, nearly the entire surface consists of solvent molecules (A) and the presence of molecules of B affects only a small number of solvent molecules. In this case, the vast majority of A interactions are with other A molecules, which means that Raoult's law applies to the solvent vapor. Solutions under these conditions are called **ideal-dilute solutions**. The **chemical potential** of the solvent in the liquid phase is then given by (see Topic D1):

$$\mu_A = \mu_A^{\ominus} + RT\ln a_A \text{ with } a_A = x_A$$

and $\mu_A^{\ominus}$ is the **standard chemical potential** of the **solvent**, when $a_A = x_A = 1$, which is the chemical potential of pure liquid A. By definition, the activity of a pure liquid is unity (see Topic C1). As $x_A = 1 - x_B$, adding a solute decreases $x_A$ below unity, and it follows that the chemical potential of an impure solvent is always less than a pure one. This means that an impure solvent is more stable than a pure one, as it has a lower molar **Gibbs free energy** (see Topic B6), so that adding a solute decreases the tendency for a solvent to vaporize or freeze. This is the origin of the **colligative properties** of the solvent.

Under these very dilute conditions the solute molecules, B, are surrounded almost entirely by molecules of A; very different conditions from those which are present in pure liquid B. However, experimentally its partial vapor pressure, $p_B$, is still found to be proportional to its mole fraction, $x_B$:

$$p_B = K_B x_B \qquad \text{when } x_B \ll 1$$

where $K_B$ is a constant (not to be confused with the base dissociation equilibrium constant, see C1). This equation is called **Henry's law** and $K_B$ is often called the **Henry's law constant**. Its value is constant for a particular solute, B, but also depends on the nature of the solvent, A, as dissolution of B involves the formation of B–A interactions and the disruption of A–A interactions. Strong B–A interactions relative to A–A interactions will tend to favor B being in the liquid and reduce $p_B$ (resulting in a small $K_B$) whilst relatively weak B–A interactions will lead to a larger $K_B$. Although Henry's law applies only to dilute solutions, many real systems such as gases dissolved in water or in blood are just such dilute solutions. In this case, knowledge of the $K_B$ values of the gas for these systems allows the mole fraction (and from this the concentration) of these gases to be determined at any partial vapor pressure or partial pressure.

# D3 COLLIGATIVE PROPERTIES

## Key Notes

| | |
|---|---|
| **Characterization using colligative properties** | A colligative property of a solution is a property that depends only on the number of solute molecules present. Measurement of one of these properties allows the determination of the molality or the concentration of the solute, from which the molecular mass of the solute can be calculated, if the mass of solute in solution and the volume of the solution are known. |
| **Depression of solvent freezing point** | When a small amount of involatile solute B is added to a solvent A to make an ideal-dilute solution, the depression of the freezing point of A, $\Delta T_f$, is related to the molality of B, $m_B$, by $\Delta T_f = K_f m_B$, where $K_f$ is the cryoscopic constant, or the freezing point depression constant. $K_f$ is a constant for a given solvent, A. |
| **Elevation of solvent boiling point** | When a small amount of involatile solute B is added to a solvent A to make an ideal-dilute solution, the elevation of the boiling point of A, $\Delta T_b$, is related to the molality of B, $m_B$, by $\Delta T_b = K_b m_B$, where $K_b$ is the ebullioscopic constant, or the boiling point elevation constant. $K_b$ is a constant for a given solvent, A. |
| **Osmotic pressure** | The osmotic pressure, $\Pi$, established between pure a solvent, A, and an ideal-dilute solution of a solute, B, across a membrane permeable only to solvent is given by $\Pi \approx c_B RT$, where $c_B$ is the concentration of solute B. Measurement of this osmotic pressure allows $c_B$ to be determined. |
| **Related topics** | Fundamentals of equilibria (C1)      Solutions (D2) <br> Non-electrolyte solutions (D1) |

**Characterization using colligative properties**

In an **ideal-dilute solution**, when an involatile solute, B, is added to a solvent, A (see Topic D2) the **chemical potential** of the liquid solvent is lowered, as $\mu_A = \mu_A^{\ominus} + RT\ln x_A$, where $x_A$ is the mole fraction of A (unity for pure A and decreasing when B is added) and $\mu^{\ominus}$ is the standard chemical potential (or molar Gibbs free energy) of pure liquid A. This effect is independent of the chemical nature of B, but merely depends upon the number of moles of B added, determined by its mole fraction, $x_B$, as $x_A = 1 - x_B$ (see Topic D1). This thermodynamic stabilization of the solvent results in measurable changes, such as the depression of the solvent vapor pressure, a decrease in the freezing temperature and an increase in the boiling point, which only depend on the number of moles of B added, and not its chemical structure. Such properties are called **colligative properties**. Measurement of one of these properties is useful when characterizing a substance whose **molecular mass** is unknown. Addition of a given mass of this substance as B to a solvent to form an ideal-dilute solution allows the determination of a colligative property to obtain $x_B$, **molality** $m_B$ or **concentration** $c_B$. Using the number of moles, volume or mass of the solvent (all of which are related) respectively then enables the determination of the number of moles of B added. This, in conjunction with the added mass of B, allows the **molar mass** (or mass of one mole) of B to be determined.

**Depression of solvent freezing point**

For the equilibrium corresponding to the physical transition of A between solid and liquid:

$$A_{(s)} \rightleftharpoons A_{(l)}$$

the activity of the solvent A in the liquid is given by $a_A = x_A$ (see Topic D1). This means that the small amount of the solute, B, added in an ideal-dilute solution affects this activity. Equilibrium is established only at the freezing temperature or freezing point (which is the same as the melting temperature or melting point) of A. Below this temperature, all of the liquid A has frozen and A is only present as solid; above this temperature, all the solid A has melted and only liquid A is present. At the equilibrium temperature, the change in Gibbs free energy and hence the equilibrium constant, $K$, for the reaction is dominated by A (as there is a minute amount of B present) and is given by $K = x_A$, as pure solid A has an **activity** of 1 (see Topic C1).

The temperature variation of this equilibrium constant is given by the **van't Hoff equation** (see Topic C1):

$$\ln K_{T1} - \ln K_{T2} = \ln\left(\frac{K_{T1}}{K_{T2}}\right) = \frac{-\Delta H_{fus}^{\ominus}}{R}\left[\frac{1}{T_1} - \frac{1}{T_2}\right]$$

where $\Delta H_{fus}^{\ominus}$ is the **standard enthalpy of fusion** (the enthalpy required to melt a mole of solid A under standard conditions). For pure liquid A, $K_{T2} = x_A = 1$ and $T_2 = T_f^0$, the freezing point of pure liquid A, whereas when B is added, $K_{T1} = x_A$ and $T_1 = T_f$, the new freezing point. Therefore:

$$\ln(x_A) = \frac{-\Delta H_{fus}^{\ominus}}{R}\left[\frac{1}{T_f} - \frac{1}{T_f^0}\right]$$

which gives

$$-x_B \approx \ln(1 - x_B) = \frac{-\Delta H_{fus}^{\ominus}}{R}\left[\frac{1}{T_f} - \frac{1}{T_f^0}\right]$$

when a small amount of B is added and $x_B$ is small compared to 1. Under these conditions, the change in the freezing temperature is also small and:

$$\frac{1}{T_f} - \frac{1}{T_f^0} = \frac{T_f^0 - T_f}{T_f^0 T_f} \approx \frac{\Delta T_f}{\left(T_f^0\right)^2} \qquad \text{where} \qquad \Delta T_f = T_f^0 - T_f$$

$\Delta T_f$ is the **freezing point depression**, the decrease in the freezing temperature of the solvent A on adding B (defined as a positive value). Thus:

$$x_B = \frac{\Delta H_{fus}^{\ominus}}{R\left(T_f^0\right)^2}\Delta T_f$$

Usually, this equation is written in terms of the **molality** of B, $m_B$ (see Topic D1). At these low concentrations, $x_B = m_B M$, where $M$ is the mass of one mole of the solvent, or the **molar mass**. The equation then becomes:

$$\Delta T_f = \frac{R\left(T_f^0\right)^2 M}{\Delta H^{\ominus}_{\text{fus}}} m_{\text{B}} = K_f m_{\text{B}} \quad \text{where} \quad K_f = \frac{R\left(T_f^0\right)^2 M}{\Delta H^{\ominus}_{\text{fus}}}$$

$K_f$ is the **cryoscopic constant** or the **freezing point depression constant** of the solvent A (not to be confused with an equilibrium constant, see section C). $K_f$ can in principle be calculated from the enthalpy of fusion and the freezing point of pure liquid A, but in practice values have been measured experimentally and tabulated for a range of solvents. This value can be used, along with the measured value of $\Delta T_f$, to determine $m_{\text{B}}$ and/or $x_{\text{B}}$ for any solute.

**Elevation of solvent boiling point**

For the equilibrium corresponding to the physical transition of A between liquid and vapor in an ideal-dilute solution:

$$\text{A}_{(l)} \rightleftharpoons \text{A}_{(g)}$$

the **activity** of the solvent A in the liquid is again given by $a_{\text{A}} = x_{\text{A}}$ (see Topic D1). The equilibrium constant for this process, which is again dominated by A, is (see Topic C1):

$$K = \frac{p_{\text{A}}}{p^{\ominus} x_{\text{A}}}$$

where $p^{\ominus}$, the standard pressure, has a value of 1 atmosphere. At the boiling point, the pressure of the solvent, $p_{\text{A}}$ is also 1 atmosphere, and so $K = 1/x_{\text{A}}$. By analogy with the freezing point equilibrium:

$$-\ln x_{\text{A}} = \frac{-\Delta H^{\ominus}_{\text{vap}}}{R} \left[ \frac{1}{T_b} - \frac{1}{T_b^0} \right]$$

where $T_b$ and $T_b^0$ are the boiling points of solvent A (with B added) and pure solvent A, respectively, and $\Delta H^{\ominus}_{\text{vap}}$ is the **standard enthalpy of vaporization** of solvent A (the enthalpy required to boil a mole of liquid).

The change in sign of the expression means that adding solute B stabilizes liquid A and leads to an increase in the boiling point of the solvent, which corresponds to a positive value of $\Delta T_b = T_b - T_b^0$, which is known as the **boiling point elevation**. The final expression is:

$$\Delta T_b = \frac{R\left(T_b^0\right)^2 M}{\Delta H^{\ominus}_{\text{vap}}} m_{\text{B}} = K_b m_{\text{B}} \quad \text{where} \quad K_b = \frac{R\left(T_b^0\right)^2 M}{\Delta H^{\ominus}_{\text{vap}}}$$

is the **ebullioscopic constant** or the **boiling point elevation constant**. As with the cryoscopic constant, this is a constant for any given solvent, and values have been measured experimentally and tabulated, which again allows the determination of $x_{\text{B}}$ and $m_{\text{B}}$ for any solute from $\Delta T_b$.

**Osmotic pressure**

**Osmosis** is the movement of a solvent from a solution of lower solute concentration (higher solvent concentration) to one of higher solute concentration (lower solvent concentration). In **osmotic pressure measurements**, a semi-permeable membrane, which is permeable only to solvent (as it has holes that are small enough to prevent large solute molecules passing through) separates two liquids. This means that this technique is only used for relatively large solute molecules, often polymers or biological macromolecules. Typically, one liquid is pure water and the other is the solute solution of interest. This produces a flow of water from the solvent to the solute solution. The experimental apparatus (*Fig. 1*) incorporates two identical vertical columns, one for each liquid, and the height of the liquid in the solution column increases relative to that in the pure solvent column due to this net flow.

The extra height (and mass of water) in the solute column compared to the solvent column produces an excess gravitational force (and pressure) in the solute compartment. This produces an opposing pressure to the osmotic pressure, trying to squeeze solvent back through the membrane. When this pressure exactly opposes the **osmotic pressure, $\Pi$**, the pressure driving osmotic solvent flow, the flow ceases and an equilibrium is established with the heights of the liquid in the two columns remaining constant. The pressure exerted at the foot of a column of solvent is proportional to its height, so a reading of the difference in heights of the two columns then leads directly to the difference in pressure, which is itself equal to $\Pi$. The osmotic pressure of an ideal-dilute solution is then given by:

$$\Pi V \approx n_B RT$$

where $n_B$ is the number of moles of solute B. Thus:

$$\Pi \approx c_B RT$$

which enables $c_B$, the concentration of solute B in the solution in the right hand compartment at equilibrium, to be measured at any temperature $T$.

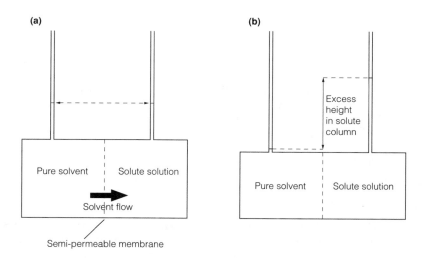

Fig. 1. Schematic diagram of the apparatus for measuring osmotic pressure. (a) Apparatus at the start of the measurement; (b) apparatus at equilibrium

# **D4** PHASE EQUILIBRIA

## Key Notes

| | |
|---|---|
| **Melting and freezing points** | The melting point and the freezing point are identical for a pure substance, but not necessarily for a mixture. The melting temperature of a pure substance increases with increasing pressure, except for water, where the melting temperature decreases with increasing pressure. |
| **Boiling point** | The boiling point of a pure substance increases with increasing pressure. This increase is relatively small compared to the increase in melting temperature with pressure. As the pressure is increased, the density of the liquid and the vapor become increasingly similar. Eventually, at the critical point, denoted by the critical temperature and pressure, the densities become equal, it is impossible to distinguish between the liquid and vapor phases, and there is no longer a phase transition. |
| **Phase diagrams of a single species** | The variation of the boiling point and freezing (melting) points can be plotted as lines on a diagram of $p$ versus $T$. This produces a phase diagram that shows the phase(s) present at any state point $p$, $T$. At one point on the diagram where the two lines meet, called the triple point, all three phases (solid, liquid and gas) coexist. There are three variables, $p$, the molar volume, $V_m$, and $T$ for the system, and any two can form the axes of a phase diagram. The number of degrees of freedom, $F$, is the minimum number of independent variables that can be varied without changing the number of phases in the system. $F = 2$ for a single-phase region, which means that a single phase can be maintained whilst moving in two dimensions. When two phases are in equilibrium, $F = 1$ and two phases can only be maintained by moving along the two-phase line in one dimension. At the triple point, there are three phases, $F = 0$ and this only occurs at this fixed point on the diagram. |
| **Related topic** | Phase diagrams of mixtures (D5) |

**Melting and freezing points**

When a pure solid species is heated through its **melting point (melting temperature)**, the solid changes to a liquid. At the **melting point**, an equilibrium is established between the solid and liquid phases. Conversely, when a pure liquid is cooled through its **freezing point** or **freezing temperature**, the liquid is transformed to a solid, and at the freezing point, equilibrium is established. Thus the melting and freezing temperatures of a pure substance are identical and in this case the terms can be used interchangeably. (However this is often not the case for a mixture, as the freezing temperature, where solid first starts to appear from a solid mixture, is often not the same as the melting temperature, where solid first starts to melt in a solid mixture (see Topic D5).)

At the melting point, the equilibrium for a pure species A is:

$$A_{(s)} \rightleftharpoons A_{(l)}$$

and as for all equilibria, the change in **Gibbs free energy** for the forward reaction, $\Delta G$, is zero under all conditions (see Topic C1). For each phase, for a small change in free energy, $dG$ (see Topic B6), $dG = Vdp - SdT$ and therefore $d\Delta G = \Delta Vdp - \Delta SdT$, where $\Delta G$, $\Delta V$ and $\Delta S$ are the changes in Gibbs free energy, volume and entropy during the forward reaction, so that at equilibrium, $\Delta G = G_{(l)} - G_{(s)} = 0$, $\Delta V = V_{(l)} - V_{(s)}$ and $\Delta S = S_{(l)} - S_{(s)}$, with $\Delta G$, $\Delta V$ and $\Delta S$ being the change in the Gibbs free energy, volume and entropy on melting respectively. This means that:

$$\frac{dp}{dT} = \frac{\Delta S}{\Delta V}$$

The change in entropy on melting is always positive, as liquid species have more freedom of movement than solid species. The change in volume is also usually positive, as melting a solid produces a liquid in which the molecules move around more (have more translational energy), and as a consequence occupy more space. In this case, $dp/dT$ is positive, and increasing the pressure increases the melting temperature. A notable exception to this is water, as solid water (ice) has an open, hydrogen-bonded structure, which occupies more volume than liquid water. This is why icebergs float, and as a consequence $\Delta V$ is negative; in this case $dp/dT$ is negative (*Fig. 1b*).

**Boiling point**

At the boiling point of a liquid A, an equilibrium is established for the physical transition of A between liquid and vapor:

$$A_{(l)} \rightleftharpoons A_{(g)}$$

Again, the equation $\dfrac{dp}{dT} = \dfrac{\Delta S}{\Delta V}$ can be applied to this equilibrium, but in this case with the changes in volume and entropy being $\Delta V = V_{(g)} - V_{(l)}$ and $\Delta S = S_{(g)} - S_{(l)}$, for the transformation between liquid and vapor. $\Delta S$ always has a positive value, as molecules have more freedom of movement in the vapor than in the liquid. Furthermore, $\Delta V$ is always positive and is usually much larger than that observed in a melting transition, as the volume occupied by a mole of gas is much larger than a mole of liquid, whereas the difference in the volumes of liquid and solid are comparatively small. As a result, $dp/dT$ is positive, but has a smaller value than for the melting transition, which means that increasing the pressure produces a smaller increase in the boiling point than the melting point.

If $dp/dT$ is changed according to this equation, thereby ensuring that equilibrium is maintained, the increase in pressure tends to compress the vapor volume, increasing its density, whilst the increase in temperature tends to weaken the liquid intermolecular forces, decreasing its density. Eventually, at the **critical point**, characterized by a **critical pressure** and a **critical temperature**, the densities of vapor and liquid become equal, the two phases are indistinguishable and there is no longer any measurable phase transition.

**Phase diagrams of a single species**

The boiling equilibrium condition is most easily represented as a line on a plot of the pressure, $p$ against the temperature, $T$ (*Fig. 1*). In this plot, this condition can be represented as a line of positive gradient, so that any point on this line corresponds to the situation where liquid and vapor are at equilibrium. Away from the line, the equilibrium condition no longer applies; above the line (at increased pressure and/or decreased temperature) only liquid exists, whereas below the line (at increased temperature and/or decreased pressure) there is

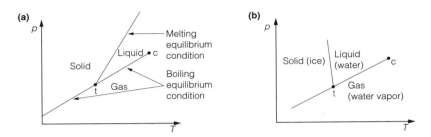

Fig. 1.   *Pressure–temperature plot (phase diagrams) resulting from plotting the boiling point and melting point equilibrium condition lines. (a) Normal plot; (b) plot for water.*

only vapor. The line terminates at c, the critical point where liquid and gas are indistinguishable.

The melting equilibrium condition line (the melting point line) will also normally be a line of positive gradient (*Fig. 1a*), and its larger value of $dp/dT$ ensures that it is always steeper than the boiling point line, which means that the two lines intersect at a point t. At any point on the melting point line, there is an equilibrium between the solid and liquid phases. Above the line, at greater pressures and/or lesser temperatures, only solid is present and below the line only liquid is to be found. The **triple point**, t, is the only point where solid, liquid and gas all exist in equilibrium and must occur at a specific pressure and temperature. For water, the melting point line is of negative gradient, which produces the plot shown in *Fig. 1b*.

These plots are known as **phase diagrams**. Solid, liquid and gas are each a different phase (see Topic D5) and so these diagrams allow prediction of the nature of the phase(s) present for any condition of $p$ and $T$ (any state point on the diagram). In fact, $p$ and $T$, which are both **intensive variables** (see Topic B1) are all that is required to specify the **state** of the system (its molecular disposition or the number and amount of the phases present and their composition, see Topic D5). This is because although the pressure, $p$, the molar volume, $V_m$, and the temperature, $T$, are the three intensive variables used to define the state of a single substance system, these are linked by an **equation of state** (for example $pV_m = RT$ for a **perfect gas**, see Topic A1) and knowledge of only two of these variables is necessary, as the equation of state can be used to calculate the third and specify the **state**. As a consequence, any two of these three variables can be used as the axes of a single substance phase diagram. In all cases, the solid phase is favoured at low volumes and temperatures and high pressures, the gas (or vapor) is favored at high temperatures and volumes and low pressures and the liquid is favored at intermediate conditions, which simplifies phase diagram labeling.

A useful concept, especially when applied to multi-component (multi-substance) systems (see Topic D5), is the **number of degrees of freedom, $F$,** of the system. This is the minimum number of intensive variables that can be varied without changing the number of phases in the system. $F = 2$ for the regions where gas, liquid or solid only are present, as changes in two variables (for example $p$ and $T$) are possible without a phase change occurring. This corresponds to being able to move in two directions on the phase diagram (*Fig. 2a*).

On the melting point or the freezing point lines, $F = 1$, as movement in one direction, up and down the line only, is possible in order to maintain the two phases at equilibrium. This means that there is only one independent variable,

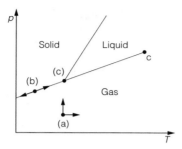

*Fig. 2. The concept of the number of degrees of freedom, F, as applied to movement on a phase diagram. (a) F = 2; (b) F = 1; (c) F = 0.*

as the two variables are related by the equation of the line (*Fig. 2b*). At the triple point, which occurs at one fixed point on the diagram, no movement is possible and $F = 0$ (*Fig. 2c*).

# D5 PHASE DIAGRAMS OF MIXTURES

---

## Key Notes

| | |
|---|---|
| **The phase rule** | The phase rule, $F = C - P + 2$, is used to determine the number of degrees of freedom in any system. $P$ is the number of phases present, each of which must be uniform chemically and physically throughout. A pure solid, a pure liquid and a pure gas are each separate phases. A mixture of gases is one phase. A mixture of two miscible liquids is one phase and of two immiscible liquids is two phases. $C$ is the number of components, which is usually equal to the number of different chemical species in the system. |
| **Phase diagrams for two components** | For a two component system consisting of a mixture of A and B, the maximum value of $F$, when $P = 1$, is $F = 3$. Thus, three variables would need to be plotted in the phase diagram. Instead, the pressure is fixed at 1 atmosphere and the remaining two variables are plotted. These are usually chosen to be temperature and the mole fraction of A, $x_A$, plotted in a temperature–composition phase diagram. |
| **Partially miscible liquids** | Two partially miscible liquids, A and B, have a temperature–composition phase diagram which either displays an upper consolute temperature, above which the two liquids are completely miscible for all compositions, or a lower consolute temperature below which they are completely miscible, or both. In the two-phase region, specifying the temperature is sufficient to calculate the composition of each phase and specifying the overall composition allows determination of the amount of each phase. |
| **Solid–liquid phase diagrams** | A mixture of two miscible liquids, A and B, has its lowest freezing point at the eutectic temperature and composition. At this point, the liquid freezes to form both solid A and solid B. At all other temperatures the liquid freezes to form first solid A or solid B, before the other solid forms at the eutectic temperature. |
| **Cooling curves** | The temperature–composition phase diagram can be used to predict the shape of the cooling curves for any mixture. In practice, cooling curves are obtained experimentally across the composition range and used to construct phase diagrams. |
| **Liquid–vapor phase diagrams** | For an ideal miscible liquid mixture of A and B, the liquid–vapor temperature–composition phase diagram can be used to determine the compositions of both vapor and liquid at any temperature in the two-phase region. The relative amounts of each phase can also be determined if the overall composition is known. When B is the more volatile component, the vapor is richer in B and liquid richer in A. Separation of pure A and B can be achieved by distillation. For some non-ideal systems, there is a low-boiling or high-boiling azeotropic point on the diagram. In this case separation of pure A and B by distillation is not possible. |

**The phase rule**

For phase diagrams of mixtures of different chemical species, the **phase rule** can be used to determine the **number of degrees of freedom, $F$,** in the system (see Topic D4). This is given by:

$$F = C - P + 2$$

where $C$ is the **number of components** and $P$ is the **number of phases** present. The number of components is the number of independent chemical species in the system. This is usually equal to the number of different chemical substances present; for example, a mixture of benzene and water would have two components. However, in a few cases, new chemical species are formed by reaction. An example is the ionization of a weak acid (see Topic C3):

$$HA_{(aq)} + H_2O_{(l)} \rightleftharpoons H_3O^+_{(aq)} + A^-_{(aq)}$$

In this case, although there appears to be four chemical species, there are two equations linking them; the **equilibrium constant** expression (see Topic C1) and an equation maintaining the overall **electroneutrality** of the system, which equates the number of cations and ions to maintain no overall charge. This means that in reality there are only two components. Generally $C = S - R$, where $C$ is the number of components, $S$ is the number of chemical species present and $R$ is the number of different equations linking them. For systems involving substances that ionize, the number of components is generally equal to the number of chemical species present without the complication of ionization. This is because ionization produces the same number of extra chemical species as equations linking them and increases both $S$ and $R$ by the same amount.

A **phase** is rigorously defined as a part of the system that is uniform both physically and chemically throughout. A pure solid, a pure liquid and a pure gas are each a phase, as in each the density and chemical composition are identical at all locations. By this criterion, a mixture of different chemical species can also be one phase, as long as the mixing is so thorough that at any location the relative amounts of all the species is the same as any other location. Mixing in gases is very efficient and there is only ever one gas phase in a mixture. A mixture of two liquids can be either two phases if the liquids are **immiscible** (do not mix and form two separate liquids) or one phase if they are completely **miscible** (each completely soluble in the other). Solid atoms and molecules are held tightly in the solid lattice and there is often an energy penalty for mixing them, so there is usually a phase for each solid. However, homogeneous one-phase mixtures can be formed from two solids consisting of atoms of molecules with comparable size and structure, such as in metal alloying.

**Phase diagrams for two components**

Using the **phase rule**, for a system of two components, $F = 4 - P$ and as the minimum number of phases that could be present in a system is one, the maximum value of $F$ is three. Three intensive variables would be required to define the state of the system, which would involve plotting and interpreting a three-dimensional phase diagram. This is complicated to reproduce on paper, so one variable is kept constant to avoid this. The variable chosen is usually the pressure, which is fixed at one atmosphere, the ambient pressure in experimental measurements. This reduces the number of degrees of freedom which remain to be plotted, $F'$, by one, so that $F' = 3 - P$ and the maximum value of $F'$, when $P = 1$, is $F' = 2$. These remaining two variables are plotted as the axes of the two-dimensional phase diagram. For a mixture of two species A and B, the two variables normally chosen are the temperature, $T$, as the y-axis and $x_A$, the mole fraction of A (which specifies the overall composition of the mixture) as the x-axis. It is not necessary to plot $x_B$, as this can be calculated from the relationship $x_B = 1 - x_A$. This results in a **temperature–composition phase diagram**.

**Partially miscible liquids**

A typical temperature–composition phase diagram observed for partially miscible liquids is shown in *Fig. 1a*.

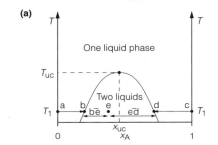

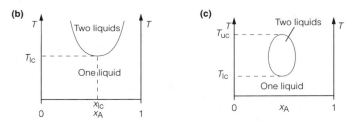

*Fig. 1.* *Temperature–composition phase diagrams for two partially miscible liquids displaying (a) an upper consolute point, (b) a lower consolute point, (c) both an upper and a lower consolute point.*

When pure liquid A is added to pure liquid B at a constant temperature $T_1$, the system moves from **state point** a to state point b. Initially, liquid A is completely soluble (miscible) in liquid B but at point b, a saturated solution of solute A in solvent B is produced. A phase transition line is drawn, as the addition of more A produces two liquid phases. Alternatively, if B is added to pure liquid A at $T_1$ (state point c to d on the diagram), the behavior will be similar. Again there is a phase transition line on the diagram at d, which corresponds to a saturated solution of B in A.

A mixture of A and B prepared at an overall mole fraction and temperature given by e separates into two liquid phases of composition b and d. Indeed, the same two saturated solutions would exist for any mixture at the same temperature, $T_1$, with an overall composition corresponding to a state point on the line between b and d in the two-phase region. This is consistent with the phase rule, as in the two-phase region when $P = 2$, $F' = 1$ and specifying the temperature is all that is required to fix the compositions of the two phases. However, the amount of each liquid would vary with the position of the state point. If the state point were to lie closer to b, then more of the liquid phase of composition, $x_A$, given by b would be present. If it were to lie closer to d, then more of the phase of composition, $x_A$, given by d would be present. The ratio of the number of moles of the liquid of composition given by b, $n_b$, to the liquid of composition given by d, $n_d$, for a mixture of overall composition given by e is therefore given by the **Lever rule**:

$$\frac{n_b}{n_d} = \frac{\overline{ed}}{\overline{be}}$$

where $\overline{\phantom{xx}}$ denotes the length of a line, so that $\overline{be}$ and $\overline{ed}$ are the lengths of the lines between b and e and between e and d, respectively.

These rules are general, and can be applied to all two-phase regions in any two component mixture. In all cases, the composition of the two phases can be found by drawing a horizontal line (a line at constant $T$) through the state point corresponding to the temperature and overall composition of the mixture. The compositions of the two phases will then be determined by the mole fractions of the points where this line intersects the boundaries of the two-phase region. The Lever rule can then be used to determine the number of moles of each phase present.

On increasing the temperature, $T$, the **miscibility (solubility)** of A and B in each other often increases. This is the case in *Fig. 1a*, demonstrated by the fact that the two lines become closer on increasing $T$, indicating that more A is required to produce the saturated solution of A in B, and more B is required to produce the saturated solution of B in A. Eventually the two curves meet, at the temperature and mole fraction denoted by $T_{uc}$ and $x_{uc}$. These are known as the **upper consolute temperature** and the **upper consolute composition** respectively. $T_{uc}$ is the temperature above which only one phase is present, as liquids A and B are completely miscible at all compositions.

In contrast, if the solubility of A and B in each other increase as $T$ decreases, then the phase diagram takes the form shown in *Fig. 1b*. Now the lines get closer together as $T$ decreases, and at $T_{lc}$ and $x_{lc}$, the curves meet at the **lower consolute temperature and lower consolute composition** (a **lower consolute point**). The lower consolute temperature is the temperature below which the liquids are completely miscible and there is only one liquid phase at all compositions.

In one special case, experimentally found for mixtures of nicotine and water, there is a range of temperature over which A and B are partially miscible. This results in both an upper consolute temperature and composition, $T_{uc}$ and $x_{uc}$, and a lower consolute temperature and composition, $T_{lc}$ and $x_{lc}$, as can be seen in *Fig. 1c*.

**Solid–liquid phase diagrams**

If two completely miscible liquids, A and B, are cooled sufficiently, then solid will start to form. When the amount of solute in the solution is small (A in B or B in A), the freezing temperature, the temperature at which this occurs, is decreased by the addition of solute (see Topic D3). This produces two lines corresponding to the change in the freezing points of A and B with $x_A$ which can be plotted on the solid–liquid phase diagram (*Fig. 2a*).

Each of these lines shows a decrease with increasing mole fraction of the solutes, B and A, from the freezing temperatures of the pure solids, $T_A^0$ and $T_B^0$, when $x_A$ is near unity and zero respectively. At the mole fraction $x_{eu}$ and temperature $T_{eu}$, the two lines meet. This point is called the **eutectic point**, and this

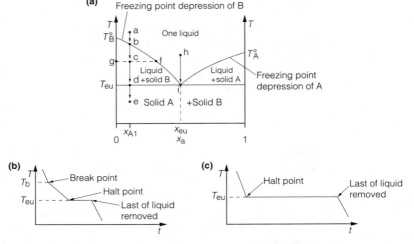

Fig. 2. *(a) Solid–liquid phase diagram for two completely miscible liquids which form two completely immiscible solids; (b) the cooling curve for $x_{A1}$; (c) the cooling curve for $x_{eu}$.*

point has a characteristic **eutectic temperature, $T_{eu}$,** and **eutectic composition, $x_{eu}$.** This corresponds to the lowest freezing temperature of the liquid A, liquid B mixture and below this temperature, only solid A and solid B are present. When heating solid A and solid B to the melting point, liquid is first produced at this temperature at all values of $x_A$; as a result, a horizontal line is drawn through $T_{eu}$, to indicate that this phase change occurs.

When a liquid mixture at the state point a is cooled, at the state point b the freezing temperature of A is reached and solid B starts to form. In the region below this line a two-phase system of solid B and liquid is present as indicated. As the system continues to cool, solid B continues to form and the liquid becomes richer in A. For example, at state point c, there is an equilibrium between solid B (state point g) and liquid (state point f). Again, the ratio of the number of moles of solid B, $n_s$, to liquid, $n_l$, is given by the **Lever rule** $n_s/n_l = \overline{cf}/\overline{gc}$.

As the temperature decreases, the liquid/two-phase boundary moves to higher $x_A$, confirming that the liquid is becoming richer in A, and from the Lever rule the relative amount of solid increases, confirming that solid B continues to be formed.

At state point d, the boundary with the solid A and solid B region is reached. Here the remaining liquid solidifies to form solid A and solid B before passing into the two-phase solid region on further cooling and on to state point e.

The same procedure can be used at high $x_A$ to show that the region bordered by the freezing point curve and the horizontal line at the temperature $T_{eu}$ must consist of the two-phase region where solid A and liquid are in equilibrium. This means that cooling past the freezing point will produce an increasing amount of solid A and a liquid richer in B until, at $T_{eu}$, solid A and solid B are formed.

**Cooling curves**

The easiest way experimentally to study the cooling of mixtures is by producing a **cooling curve**. From the example above (*Fig. 2a*), the cooling curve is produced by removing energy from the system at a composition $x_{A1}$, cooling the liquid from its state point a at a constant rate, whilst monitoring the change in temperature, $T$, of the system with time, $t$ (*Fig. 2b*).

Initially, simple cooling of the liquid occurs and the temperature decreases linearly with time. At the temperature $T_b$, corresponding to state point b, solid B will start to form. The formation of intermolecular bonds in solid B is **exothermic** and will give out heat, slowing the cooling rate of the system and leading to a break in gradient of the curve, or a **break point**. As the system is cooled further through state point c, solid B continues to be produced and the cooling continues to be slowed but at the point d and the temperature $T_{eu}$, the boundary with the two-phase solid A and solid B region, the first solid A starts to form. This means that there are three phases at equilibrium, and from the phase rule, $F' = 3 - P = 0$; as a consequence, the temperature cannot vary without first changing the number of phases present, and so the temperature decrease stops. This is called a **halt point**. During this time, energy is still being removed from the system and the liquid is being steadily converted to solid A and solid B at a constant rate, but the heat removed from the system is exactly balanced by the heat given out by the solidification process. The halt point lasts until all the liquid has been converted to the solids A and B, which means that the length of the halt point is proportional to the amount of liquid that remained when the system reached the solid A and solid B boundary, given by the **Lever rule**. Then, as the number of phases, $P$, is reduced to 2 again, and as no more energy can be evolved by the solidification process, the cooling of the system is resumed.

In contrast, cooling a liquid system at the eutectic composition, $x_{eu}$, produces a very different cooling curve (*Fig. 2c*). On cooling the liquid from state point h, the system does not enter either two-phase region where solid and liquid are present, and therefore it is the only composition on the diagram to not show a break point. At state point i, the first solid A and solid B are formed, and three phases (solid A, solid B and liquid) are present. This again results in a halt point, but this is the longest possible halt point, as all the liquid must be converted to solid A and solid B before cooling can recommence.

Thus phase diagrams can be used to predict cooling curves at any composition. In fact, the reverse is also true, and experimental cooling curves, measured across the composition range, are the usual method by which experimental phase diagrams are constructed. This method is convenient and general, as the rules that govern the appearance of break and halt points apply to all phases and phase transitions.

**Liquid–vapor phase diagrams**

When a completely miscible liquid mixture that is also an **ideal solution** (see Topic D1) is heated until it vaporizes, a liquid–vapor phase diagram can be constructed. For all mixtures, the boiling points of pure A and pure B will be different. If the boiling point of pure A is greater than that of pure B, at any composition, more B will be in the vapor phase than A. Both vapor composition and liquid composition lines are drawn, leading to a phase diagram with information concerning the compositions of both liquid and vapor (*Fig. 3*).

As expected, the two lines meet on the phase diagram when $x_A = 0$ at $T = T_B^0$, the boiling point of pure B and at $x_A = 1$ at $T_A^0$, the boiling point of pure A. Any mixture at a state point a, above these two lines will consist entirely of gas (vapor). Any mixture at state point b, below the two lines, will consist entirely of liquid. However, for any mixture with a state point, c, between the lines, two phases, both liquid and vapor, will be present. The composition of the two

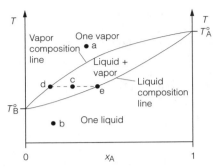

Fig. 3.   Temperature–composition plot for the boiling point variation of an ideal solution, plotted with regard to both compositions.

phases can again be determined by drawing a horizontal line through c and determining the points at which this intersects the boundaries of this two-phase region, with the vapor and liquid compositions given by the mole fractions at state points d and e respectively and the number of moles of vapor, $n_g$, relative to liquid, $n_l$, again given by the **Lever rule** $n_g/n_s = \overline{ce}/\overline{dc}$. It should be noted that metal alloys are often completely miscible **ideal solid solutions** of two metals, and so their solid–liquid phase diagrams are often similar in form to this diagram. In this case it is a liquid mixture rather than a vapor mixture which is the phase seen at high temperature in the phase diagram (*Fig. 3*), and a solid mixture rather than a liquid mixture which is the phase at low temperature. A mixture with a state point such as c is then in the two-phase (solid and liquid) region, and d and e give the composition of the liquid and solid respectively.

The liquid–vapor phase diagram has a practical use in determining the length of column required in the separation by **distillation** of two liquids, A and B (*Fig. 4*). This is often carried out on a still consisting of a heated vessel containing the liquid mixture, above which is a column containing glass beads or glass rings. Heating the liquid to boiling, state point a, at the foot of the column, produces a vapor at state point b. This then rises up the column and condenses on a glass bead to give liquid of the same composition, state point c. It is then immediately vaporized, giving a vapor at state point d that rises further up the column to condense on another glass bead as a liquid at state point e. Immediate vaporization then gives a vapor at state point f, which again rises to condense further up the column. Each successive vaporization and condensation results in a vapor which is richer in the more volatile component, B, and is termed a **theoretical**

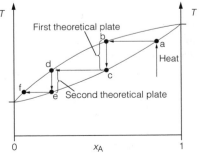

Fig. 4.   Use of the liquid–vapor phase diagram to calculate the number of theoretical plates required in distillation.

**plate**. The number of theoretical plates is proportional to the length of the column and if the still contains a large enough number of theoretical plates, the vapor eventually consists of pure B, which can be removed from the top of the column as a liquid by condensation. Continual removal of pure B by distillation leaves the liquid richer in A and the boiling point increases, until when the boiling point is $T_A^0$, only pure A remains, and A and B have been separated.

When the mixture of A and B is not ideal, marked differences are seen in the phase diagram. This is especially the case when there are significant differences in interactions between molecules of A, molecules of B and molecules of B and A. For example, when the interactions between B and A are much less than those between A and A, and between B and B, the boiling point is lowest when both B and A are present. The phase diagram for this system (*Fig. 5a*) shows the lowest boiling point at the **azeotropic temperature, $T_{az}$,** and the **azeotropic composition, $x_{az}$,** known as the **azeotropic point**. The system is then called a **low-boiling azeotrope**. At this point the composition of the vapor and the liquid are the same, and so the vapor and liquid boiling point curves meet. This has important consequences when distilling these mixtures. As before, when the particular liquid mixture shown is heated to state point a, the condensed vapor in the column has a composition c at the first theoretical plate, e at second and so on. However, eventually the composition of the condensed vapor reaches the azeotropic composition, $x_{az}$, at which point the composition of condensed vapor and vapor remains the same and no further change in composition of either is possible. This means that separation of both pure A and pure B is not possible for this system, as the vapor will always condense at the top of the column with the azeotropic composition. A similar argument applies to those mixtures rich in A, which also produce condensate at the azeotropic composition.

Alternatively, the interaction between A and B is often larger than between molecules of A and molecules of B. In this case, the boiling point is higher than the mixture and a **high-boiling azeotrope** is seen in the phase diagram (*Fig. 5b*) at the **azeotropic temperature, $T_{az}$,** and the **azeotropic composition, $x_{az}$.**

Distillation of a liquid mixture, by heating to state point a will then produce condensed vapor of composition c, e ... at successive theoretical plates, eventually leading to pure B as the condensate. However, continual removal of this condensate will lead to a liquid progressively richer in A in the distillation vessel, until the liquid reaches the azeotropic composition. At this point, both liquid and vapor have the azeotropic composition and no further separation is possible. A similar

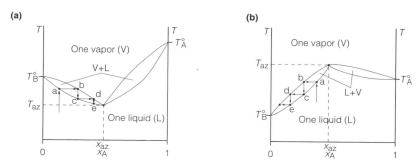

*Fig. 5.    Phase diagram for a liquid mixture that (a) shows a low-boiling azeotropic point, (b) shows a high-boiling azeotropic point.*

argument applies to those liquid mixtures rich in A. In this case distillation initally produces pure A as the condensate, but as a consequence the liquid becomes progressively richer in B, and when the liquid reaches the azeotropic composition, no further separation is possible.

# E1 IONS IN AQUEOUS SOLUTION

## Key Notes

| | |
|---|---|
| **Ionic models** | Ions are charged chemical species that at low concentrations are stabilized by an energetically favorable interaction with water. A hydration shell of coordinated water molecules is formed that increases the size of the hydrated ion. The general terms for any solvent are solvation shell and solvated ion respectively and both specific and general terms are used for water. |
| **Thermodynamic properties** | Ion hydration is an exothermic process. The addition of ions to a solution decreases the solvent volume and solvent entropy in the solvation shell but increases solvent volume and entropy in the zone between the solvation shell and bulk solution. For small ions, the former effect dominates the latter, there is an overall decrease in volume and entropy and the ion is termed structure-making. For larger ions, termed structure-breaking, the opposite is true and there is an increase in volume and entropy. |
| **Qualitative treatment of ionic interaction** | As the concentration of ions in solution increases, electrostatic ion–ion interactions become more important. An ionic atmosphere of oppositely charged counterions forms around each ion, further stabilizing the system. This clustering affects the ion thermodynamics. |
| **Related topics** | Enthalpy (B2)  Weak intermolecular interactions Free energy (B6)  (H6) Thermodynamics of ions in solution (E2) |

**Ionic models**

Ions are charged chemical species. Positively charged ions are called **cations** as they travel to the cathode of an **electrolytic cell** (see Topic E3). Negatively charged ions are termed **anions** as they travel to the anode. A **salt** is the solid which dissolves to produce these cations and anions. When a salt (**solute**) is dispersed in water (**solvent**) (see Topic D1) to form an aqueous solution of ions, also termed an **ionic** or **electrolyte solution**, there is an energetically favorable **ion–dipole interaction** (*Fig. 1*).

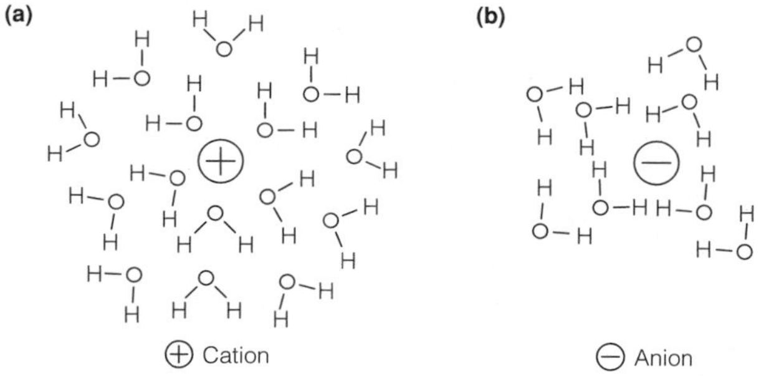

Fig. 1.   *The interaction between water and a cation and water and an anion in solution.*

Due to the different affinities of oxygen and hydrogen for electrons, the oxygen has a net fractional negative charge ($\delta-$) and the protons have a net fractional positive charge ($\delta+$), producing a **dipole** in the water molecule (see Topic H6). The charge on the cation produces a positive **electric field** (positive gradient of potential with respect to distance) in solution that aligns the water dipoles locally with oxygen closest to the ion. Conversely the anion produces a negative field that orients the water dipole with the protons closest to the ion. Both of these interactions are energetically favorable and stabilize the ion in solution. The field decreases with distance from the ion, but is strong enough to cause water molecules to cling to the surface of the ion as it moves (see Topic E7). This process is termed **hydration (solvation** generally, when solvents other than water are used) and results in the formation of a **hydration shell** (generally **solvation shell**) or a coating of water molecules, which increase the effective size of the ion in solution. The water molecules closest to the ion are held the tightest. At greater distances as the field decreases and the ion–dipole attractive force becomes comparable with the thermal force of the water molecules moving in the free liquid, there is a dynamic equilibrium, with water molecules escaping to and being replaced by molecules captured from bulk solution. At still greater distances where negligible field and hence little ion–dipole interaction remains, there is no ordering of the water molecules around the ion. Hence this water does not move with the ion and is not part of the solvation shell. The size of the solvation shell depends upon the electric field strength at the surface of the ion, $E$, and $E \propto q/r^2$, where $q$ is the charge on the ion and $r$ is the ionic radius before solvation. This means that the smallest, most highly charged ions (such as $Li^+$, $Al^{3+}$ and $F^-$) have the largest solvation shells. The overall radius of the solvated ion is the sum of the ionic radius plus the solvation shell radius, which means that the solvated ion radius is typically much larger than the ionic radius in the gas phase or in a crystal lattice and often that the smallest unsolvated ions have the largest radii when solvated.

**Thermodynamic properties**

Salts dissolve producing both cations and anions, which means that it is impossible to measure thermodynamic data for individual ion types. The solvation of ions from the gas phase (**the enthalpy of solvation**, see Topic B2) is a thermodynamically favorable **exothermic** process, which increases the stability of the ions in solution and gives out energy. However, in order for the **enthalpy of solution** ($\Delta H_{sol}$, see Topic B2) to be exothermic, this enthalpy of solution must be greater than the **lattice enthalpy**; otherwise the overall reaction is **endothermic**.

The **entropy change** due to breaking up the salt into its constituent gaseous ions is also positive, but in addition there is an entropy term due to ion solvation. Once the ion is solvated, the water molecules in the solvation shell are arranged around each ion with a center of symmetry in the system (*Fig. 1*). They are relatively tightly packed with respect to liquid water and hence have lower entropy and occupy a lower volume. In contrast, in the bulk solution, the water molecules are extensively hydrogen bonded (see Topic H6) in a relatively open tetrahedral arrangement (*Fig. 2*).

This produces a zone of water between these two regions in which the conflict in symmetry disrupts the water structure and leads to an increase in entropy (see Topic B4) and in the volume that the molecules occupy in this zone. Small, highly charged ions such as $Li^+$, $Al^{3+}$ and $F^-$ have large solvation shells compared with the size of this zone and the effects of the solvation shell dominate. They are termed **structure-making** ions as solvation of these ions leads to

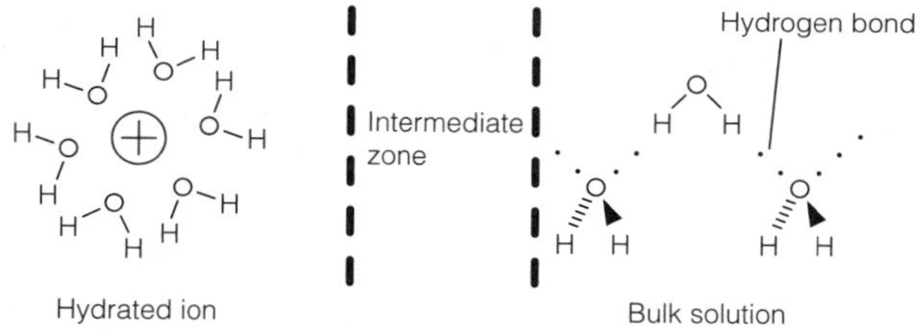

Fig. 2. *The conflict in symmetry between bulk water and water in a hydration shell.*

an overall decrease in the entropy of the system on solvation. The overall volume of the system also decreases compared to the total combined volume of the salt and water before addition. For very small or highly charged ions this volume change can be sufficiently great that the volume of the solution is smaller than the original volume of water, even though salt has been added. In contrast, for relatively large ions (e.g. organic anions and cations, $ClO_4^-$, $Rb^+$) the effects of the intermediate zone outweigh those of the solvation shell, there is an overall increase in entropy and in the total volume of the system and as a result these are termed **structure-breaking** ions.

For salt dissolution to be a **spontaneous process**, the change in **Gibbs free energy**, $\Delta G$ (see Topic B6) of the dissolution process must be negative. Thus, endothermic salt dissolution is a spontaneous process if $T\Delta S$ is sufficiently positive (see Topic B6, *Table 1*).

**Qualitative treatment of ionic interaction**

At very low ionic concentrations, the solvated cations and anions are so far apart that they do not interact significantly. At higher concentrations the ions are closer together and **ion–ion** interactions between the cations and anions are important. These are energetically favorable, as the ions tend to cluster around other ions of opposite charge (**counterions**) and avoid ions of like charge (**coions**). This process further decreases the energy of the system and stabilizes the solution. This means that although ions move dynamically in solution, on average each cation has as nearest neighbors more anions than cations and each anion more cations than anions. An **ionic atmosphere** of ions of overall opposite charge therefore surrounds each ion (*Fig. 3*).

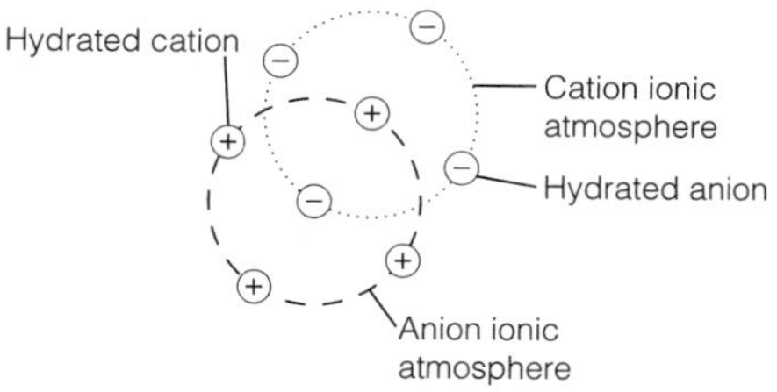

Fig. 3. *The ionic atmosphere surrounding a cation (...) and an anion (– – –) in solution.*

The effect of the ionic atmosphere increases in importance as the concentration increases and hence the stability of the ions in solution increases with increasing concentration. A good example is an equilibrium involving salt dissolution, where the addition of **inert ions** to the solution, which do not react with the cation $M^+$ or the anion $X^-$, nevertheless stabilize the solvated ions and shift the equilibrium position to favor ion dissociation (see Topic C5). This **inert ion effect** is accounted for by using the **activity** of ions, rather than their concentration to calculate thermodynamic data such as equilibrium constants.

# E2 THERMODYNAMICS OF IONS IN SOLUTION

## Key Notes

**Ionic activity**

The activity, $a_i$, of an ion, $i$, is given by:

$$a_i = \frac{\gamma_i c_i}{c^\ominus}$$

The increasing effect of the ionic atmosphere as the concentration of the ion, $c_i$ is increased is reflected in the changing value of the activity coefficient, $\gamma_i$.

**Ionic strength**

The ionic strength, $I$,

$$I = \frac{1}{2}\sum_i c_i z_i^2$$

measures the overall degree of electrostatic interaction between the ions at any concentration. It involves adding the contribution of each ion, $i$, with concentration, $c_i$, and charge $z_i$. Multiply-charged ions promote a larger degree of interaction than singly-charged ions.

**Calculation of activity coefficients**

Activity coefficients can be calculated from the Debye-Hückel limiting law (at lower $I$), Debye-Hückel law (at intermediate $I$) or the Debye-Hückel extended law (at higher $I$), as appropriate. These calculate the activity coefficient from the charge on the ion and the ionic strength and are accurate up to around $I = 1$ mol dm$^{-3}$.

**Related topics**

Free energy (B6)
Fundamentals of equilibria (C1)

Non-electrolyte solutions (D1)
Electrochemistry and ion
concentration (E5)

**Ionic activity**

The **activity**, $a_i$, of an ion $i$ (see Topic C1) is given by:

$$a_i = \frac{\gamma_i c_i}{c^\ominus}$$

where $c_i$ is the concentration of the ion in mol dm$^{-3}$, $c^\ominus$ is the standard concentration of 1 mol dm$^{-3}$ and $\gamma_i$ is the **activity coefficient** of the ion. The activity coefficient quantifies the effect of ionic interaction; at very high dilution, no ionic interaction occurs, $\gamma_i \rightarrow 1$ and ionic concentration is sufficient to calculate the **chemical potential** or **molar free energy** of the ion (see Topics B6, C1 and D1). At higher ionic concentrations, ionic interactions become significant, $\gamma_i$ is typically less than unity and electrostatic interactions are important in determining thermodynamic data.

**Ionic strength**

The overall degree of electrostatic ionic interactions is measured by using the **ionic strength** of the solution, $I$. This is given by:

$$I = \frac{1}{2}\sum_i c_i z_i^2$$

and involves adding the contribution of every ion in solution, $i$, using the concentration of the ion $c_i$ and the **formal charge** or **charge number**, $z_i$. The charge number for a general cation $M^{x+}$ is $x$ and for a general anion $X^{y-}$ is $-y$. For a 1:1 electrolyte where $x = 1$ and $y = 1$, $I$ is equal to the solution concentration, $c$. For electrolytes where $x$ and/or $y$ are greater than 1, $I > c$, which reflects the greater electrostatic interaction of ions of higher charge with other ions.

**Calculation of activity coefficients**

When the electrostatic interactions between ions and counterions are relatively weak compared with their thermal energy, $k_B T$ (i.e. at low dilution), the distribution of ions and potential at any distance from any ion can be calculated. The effect on the energy of the system of the electrostatic interactions that result can then be quantified. This procedure results in the **Debye-Hückel law**:

$$\log_{10}\gamma_\pm = \frac{-A|z_+ z_-|\sqrt{\dfrac{I}{I^\ominus}}}{1 + B\sqrt{\dfrac{I}{I^\ominus}}}$$

which can generally be applied to solutions of $I \le 0.01$ mol dm$^{-3}$ (for which this assumption of relatively weak electrostatic interactions holds). $A$ is a constant for a given solvent (0.509 for water at 298 K), $B$ is a constant for a given ion in a given solvent (conveniently $B$ is often approximately equal to 1 for most ions in water) and $I^\ominus$ is the standard ionic strength (defined as being equal to 1 mol dm$^{-3}$). This allows the **mean activity coefficient**, $\gamma_\pm$, to be calculated for a salt $M_y X_x$, where

$$\gamma_\pm = \left(\gamma_M^y \gamma_X^x\right)^{\frac{1}{x+y}}$$

This expression simply emphasizes that changes in the activity coefficient of the cation cannot be calculated without changes in the activity coefficient of the anion, as anions cluster around cations and cations cluster around anions. If it is assumed that the effects of clustering on the two activity coefficients are approximately equal (as well as the parameter $B$ for both ions), then the activity coefficient for an individual ion is given by:

$$\log_{10}\gamma_i = \frac{-Az_i^2\sqrt{\dfrac{I}{I^\ominus}}}{1 + B\sqrt{\dfrac{I}{I^\ominus}}}$$

At very low dilution (typically when $I < 0.001$ mol dm$^{-3}$), the first term in the denominator of the equation dominates the second term, and produces the simplified **Debye-Hückel limiting law** (which removes the ion dependent parameter $B$ from the expression):

$$\log_{10}\gamma_\pm = -A|z_+ z_-|\sqrt{\frac{I}{I^\ominus}} \quad \text{or} \quad \log_{10}\gamma_i = -Az_i^2\sqrt{\frac{I}{I^\ominus}}$$

At higher values of $I < 1$ mol dm$^{-3}$, the assumption of weak interactions between ions made in deriving the Debye-Hückel expression becomes increasingly untenable. Activity coefficients in this regime generally fit the expression:

$$\log_{10} \gamma_i = \frac{-Az_i^2 \sqrt{\dfrac{I}{I^\ominus}}}{1 + B\sqrt{\dfrac{I}{I^\ominus}}} + C|z_+ z_-|\frac{I}{I^\ominus}$$

where $C$ is an empirical parameter which can be adjusted to fit the data. This extra term is introduced to take account of the increasing importance of short-range ion–ion and ion–solvent forces and this equation is often called the **Debye-Hückel extended law**. The applicability of these equations to calculating mean activity coefficients is demonstrated in *Fig. 1*, where calculated values can be seen to fit experimentally determined values (see Topic E5) closely. For $I > 1$ mol dm$^{-3}$, where short-range interactions dominate and ions have increasingly incomplete solvation, theoretical calculation of activity coefficient data is notoriously unreliable.

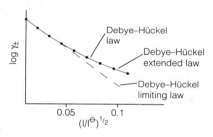

*Fig. 1. Typical experimental data for mean activity coefficients fitted to the appropriate Debye-Hückel expression.*

# E3 ELECTROCHEMICAL CELLS

---

## Key Notes

| | |
|---|---|
| **Cells** | An electrochemical cell is constructed of two half-cells, each of which is connected by an outside electrical circuit (as an electrical connection) and a salt bridge (as an ionic connection). Each half-cell consists of a metal (an electrode) at which a redox reaction occurs which establishes a characteristic potential on the metal surface. The simplest half-cell is a metal dipping into a solution of its ions. |
| **Galvanic and electrolytic cells** | In a galvanic cell, current is allowed to pass in the spontaneous direction (electrons move from low to high potential) by simply connecting the two electrodes together. For thermodynamic measurements, the potential difference (voltage) is measured without current flow by connecting a high impedance voltmeter between the electrodes. In the electrolytic cell a voltage is applied to the electrodes to force reaction in the non-spontaneous direction. |
| **Other half-cell reactions** | Other important half-cells are the metal-insoluble salt electrode, the gas electrode and the redox electrode. These last two electrode reactions do not involve a metal and an inert platinum metal electrode is required to produce the electrode potential. |
| **The standard state** | Cell voltages and half-cell potentials are often compared under standard conditions called the standard state at a specified temperature (usually 298 K). The standard state requires all reagents to be pure with all gases at unit fugacity, all ions at unit activity and all electrical connections between half-cells to be made with platinum. The biological standard state is often used for biological systems, as the standard state is inappropriate. |
| **Related topics** | Enthalpy (B2)  Electrochemical thermodynamics (E4)  Thermochemistry (B3)  Electrochemistry and ion concentration (E5)  Entropy and change (B5)  Thermodynamics of ions in solution (E2) |

---

**Cells**

A **redox** reaction is any reaction between two species where electrons are transferred from one species to another. The species losing electrons is said to be **oxidized**; as it is giving electrons away and reducing the other reactant it is also called a **reducing agent** or **reductant**. The species gaining electrons is **reduced** and acts as an **oxidizing agent** or **oxidant**, as it causes oxidation. In **electrochemistry**, this transfer of electrons between chemical species is accomplished via an external electrical circuit by using an **electrochemical cell**. This consists of two **half-cells** connected together. The simplest half-cell consists of a metal M (or **electrode**) dipping into an aqueous solution of its metal ion, $M^{z+}$ (the **electrolyte**), which has the **half-cell reaction**:

$$M^{z+}_{(aq)} + ze^- \rightleftharpoons M_{(s)}$$

The electrons in this reaction are deposited on the metal and their concentration and energy, controlled by the relative stability of the metal and metal ion, determines the electrode potential. A **salt bridge**, usually consisting of a solution or gel containing saturated KCl in a glass tube, connects the two half-cell solutions (*Fig. 1a*). An alternative method of separating the two half-cell solutions is a porous **glass frit** or porous ceramic (*Fig. 1b*), which also allows the passage of ions without mixing the solutions and enables the half cells and solutions to be combined into one container (see Topic E5). This is experimentally simpler, but introduces an extra cell voltage due to the liquid junction in the frit, and is thus avoided when making thermodynamic measurements.

**Galvanic and electrolytic cells**

The difference in potential of the two metals results in a **potential difference** (also called a **voltage** or **electromotive force**, **emf**) between the two half-cells. This can be measured by means of a high impedance voltmeter (*Fig. 1b*), which measures the voltage or driving force for reaction without allowing current to flow, from which can be calculated thermodynamic data (see Topic E4). Alternatively the reaction can be allowed to proceed by connecting the two half-cells by an outside circuit (a wire or a resistor) and allowing the current to flow. These are both examples of **galvanic cells**, where the spontaneous chemical reaction occurs (see Topic B5). Electrons flow from the electrode with the most negative potential (the **anode**, where oxidation occurs) to that with the most positive potential (the **cathode**, where reduction occurs). The salt bridge (or porous glass frit) allows ions to transfer into each half-cell. This flow counteracts the imbalance of charge that would develop in each half-cell as electrons pass from one electrode to the other, which would inhibit the reaction. The need for a salt bridge or frit is avoided if both half-cells can share a common electrolyte. This is a special case, where all redox active ions in the solution react specifically at one half-cell electrode only and therefore do not have to be separated from the other electrode.

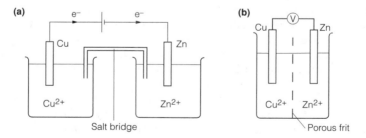

Fig. 1. *Examples of (a) an electrolytic cell incorporating a salt bridge; (b) a galvanic cell incorporating a porous frit.*

An alternative is to drive the reaction in the reverse or non-spontaneous direction. This is achieved by connecting the electrodes to an external voltage supply and applying a voltage to reverse the driving force for the flow of electrons, forcing them to move in the opposite direction (*Fig. 1a*). This is often the method by which reactive or relatively unstable species are made. This is called an **electrolytic cell** and by definition the anode in a galvanic cell is the cathode in an electrolytic cell and vice versa. Again, transfer of ions from the salt bridge or glass frit maintains electrical neutrality in the half-cells during reaction.

**Other half-cell reactions**

The **metal–insoluble salt** electrode consists of a metal M coated with a porous insoluble salt MX in a solution of $X^-$. A good example is the silver/silver chloride electrode (*Fig. 2a*) for which the half-cell reaction is $AgCl_{(s)} + e^- \rightleftharpoons Ag_{(s)} + Cl^-_{(aq)}$, where the reduction of solid silver chloride produces solid silver and releases chloride ion into solution.

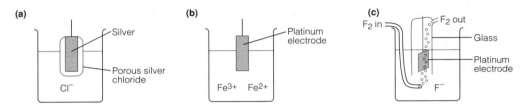

Fig. 2.   (a) The silver/silver chloride half-cell; (b) the ferric (Fe³⁺)/ferrous (Fe²⁺) half-cell; (c) the fluorine/fluoride ion half-cell.

Half-cell reactions often involve soluble reagents and not metals and these are often called **redox electrodes**. An example is the ferric/ferrous ion redox electrode (*Fig. 2b*), $Fe^{3+}_{(aq)} + e^- \rightleftharpoons Fe^{2+}_{(aq)}$. In these cases a platinum electrode is used as the metal electrode. The platinum acts at a site for the exchange of electrons between the reductant and the oxidant, resulting in electrode potential being established. It is also inert and does not form part of any redox reaction.

When one of the reagents is a gas (a **gas electrode**), for example the fluorine/fluoride gas electrode (*Fig. 2c*), $F_{2(g)} + 2e^- \rightleftharpoons 2F^-_{(aq)}$, the inert platinum electrode is also used as a site for the electron exchange, with the gas being bubbled through the aqueous solution containing the redox active ion and across the electrode surface.

**The standard state**

The value of the cell voltage, $E_{cell}$, depends upon the position of the equilibrium for each half-cell reaction, which depends upon the **activities** of ions in solution (see Topic E2), the **fugacity** (or pressure) of gas in a gas electrode and the temperature. In order to compare different cells, cell voltage values, $E^\ominus_{cell}$, are often obtained at the **standard state** (see Topic B3). For biological systems, the **biological standard state** is used, symbol $^\oplus$ (see Topic B2). In electrochemical cells, there is the additional constraint for both that the inert electrodes and electrical connections are made of platinum. $E^\oplus_{cell}$ and $E^\ominus_{cell}$ values are identical for those half-cells for which hydrogen or hydroxide ions are not involved in the cell reaction. However, there are significant differences when they are (see Topic E5).

# E4 ELECTROCHEMICAL THERMODYNAMICS

---

## Key Notes

**Notation**

A common notation is used for electrochemical cells. The symbols |, ‖ and : are used to denote a phase boundary, salt bridge and liquid junction respectively. All outside electrical connections are assumed to be present and activities (or concentrations) and fugacities (or pressures) are specified where necessary. The left hand (LH) half-cell is shown on the left-hand side of the diagram with the right-hand (RH) half-cell on the right-hand side. The cell voltage is the potential difference between RH and LH half-cells when a salt bridge is used.

**Formal cell reaction**

The formal cell reaction is obtained by taking the difference between the RH and LH half-cell reduction reactions, whilst ensuring that the same number of electrons is present in each half-cell reaction.

**Reference electrodes**

Changes in half-cell potential can be measured by measuring the voltage of a cell with the half-cell as the RH electrode and a reference electrode as the LH electrode. This is because the reference electrode maintains a constant half-cell potential even with small changes in reagent activity.

**Standard reduction potentials**

The standard reduction potential of a half-cell is the standard cell potential of a cell with the half-cell as the RH electrode and the standard hydrogen electrode as the LH electrode. $E^{\ominus}$ values, when tabulated, give an electrochemical series that shows an increase in the oxidizing power of the oxidant and a decrease in the reducing power of the reductant in the redox reaction as $E^{\ominus}$ increases. This can be used to predict whether a redox reaction between an oxidant and a reductant is spontaneous under standard conditions.

**Thermodynamic data**

The cell voltage and changes in the cell voltage with a small change in temperature at constant pressure can be used to determine changes in Gibbs free energy, entropy and enthalpy and the equilibrium constant for the formal cell reaction and the half-cell reduction reactions. Changes in the cell voltage with a small change in pressure at constant temperature can be used to calculate changes in volume for the cell reaction and half-cell reaction respectively.

**Related topics**

Free energy (B6)
Fundamentals of equilibria (C1)

Electrochemical cells (E3)
Electrochemistry and ion concentration (E5)

---

**Notation**

To avoid repetitive drawing of complicated cell diagrams, a common notation has been adopted for cells. All electrical contacts between half-cells are assumed, changes of phase (see Topic D4) are denoted by |, a salt bridge is shown as ‖ and a junction between two different solutions (a **liquid junction** in a **glass frit**) by ⋮. If there are multiple species in the same phase these are separated by commas. The cell is shown starting at the left-hand electrode and moving to the right-hand electrode through the solutions. Examples of cells under standard conditions are:

$$Pt \mid H_2(g, p = 1\ atm) \mid H^+(aq, a = 1) \parallel Cl^-(aq, a = 1) \mid AgCl(s) \mid Ag(s)$$

$$Pt \mid Fe^{3+}(aq, a = 1), Fe^{2+}(aq, a = 1) \parallel Zn^{2+}(aq, a = 1) \mid Zn(s)$$

Activity, $a$, or concentration, $c$, and pressure, $p$, or fugacity values are not necessary and generally not quoted for standard cells (as they are defined), but are important away from the **standard state** (see Topic E5). Ions in the solution that take no part in the redox reaction are generally not included. In these cells, the half-cell on the left both experimentally and as written is called the **left-hand (LH) electrode**, while that on the right is called the **right-hand (RH) electrode**. Measured values of $E_{cell}$ or $E_{cell}^{\ominus}$ are reported by convention as a positive or negative value, denoting that the right-hand (RH) electrode has a more positive or negative potential than the left-hand (LH) electrode:

$$E_{cell}^{\ominus} = E_{RH}^{\ominus} - E_{LH}^{\ominus} \text{ and } E_{cell} = E_{RH} - E_{LH}$$

For those cells with half-cells separated by a glass frit or porous ceramic, there is a small extra voltage associated with the liquid junction, which forms an extra component of $E_{cell}^{\ominus}$. This complication is avoided by using a **salt bridge**, which has a negligible voltage as it contains two liquid junctions whose potential differences cancel.

**Formal cell reaction**

The overall **formal cell reaction** is obtained by writing both half-cell reactions as reductions, e.g.

$$Pt \mid Fe^{3+}(aq), Fe^{2+}(aq) \parallel Zn^{2+}(aq) \mid Zn(s)$$

RH      $Zn^{2+}_{(aq)} + 2e^- \rightarrow Zn_{(s)}$

LH      $Fe^{3+}_{(aq)} + e^- \rightarrow Fe^{2+}_{(aq)}$

The number of electrons in each equation is then made equal (if necessary):

RH      $Zn^{2+}_{(aq)} + 2e^- \rightarrow Zn_{(s)}$

2LH      $2Fe^{3+}_{(aq)} + 2e^- \rightarrow 2Fe^{2+}_{(aq)}$

and, as with the half-cell potentials, the LH reaction is subtracted from the RH to give the formal cell reaction. Minus the LH reaction is equivalent to reversing the reaction (writing it as an oxidation):

RH      $Zn^{2+}_{(aq)} + 2e^- \rightarrow Zn_{(s)}$

–2LH      $2Fe^{2+}_{(aq)} \rightarrow 2Fe^{3+}_{(aq)} + 2e^-$

which is followed by combining species on the same side of the reaction arrows together and cancelling those which appear on both sides to give:

$$2Fe^{2+}_{(aq)} + Zn^{2+}_{(aq)} \rightarrow Zn_{(s)} + 2Fe^{3+}_{(aq)}$$

which is the formal cell reaction.

This should always result in the total charge on both sides of the reaction being equal and the removal of electrons from the equation. Balancing of this equation can also be achieved by reducing each half-cell reaction to a one-electron reduction. This results in the same cell reaction, but with the amount of reaction halved. It is important to remember this when calculating thermodynamic parameters for the formal cell reaction, as these are for the formal cell reaction as derived.

**Reference electrodes**

It is not possible to measure the potential of one half-cell, only the voltage between two half-cells (see Topic E3). However, changes in the potential of one half-cell (RH) can be measured by measuring cell potentials and choosing an LH half-cell which maintains a constant potential despite small changes in the amounts of its redox reagents. This is termed a **reference electrode**. Examples are given in *Table 1*.

*Table 1.    Examples of reference electrodes*

| Electrode | Half-cell | Reaction |
|---|---|---|
| Silver/silver chloride | $Ag(s) \mid AgCl(s) \mid Cl^-(aq, c = 0.1\ M)$ | $AgCl_{(s)} + e^- \rightleftharpoons Ag_{(s)} + Cl^-_{(aq)}$ |
| Calomel (mercury/mercurous chloride)[a] | $Hg(l) \mid Hg_2Cl_2(s) \mid Cl^-(aq)$ | $\tfrac{1}{2}Hg_2Cl_{2(s)} + e^- \rightleftharpoons Hg_{(l)} + Cl^-_{(aq)}$ |
| Hydrogen electrode[b] | $Pt \mid H_2\ (g) \mid H^+(aq)$ | $H^+_{(aq)} + e^- \rightleftharpoons \tfrac{1}{2}H_{2(g)}$ |

[a]Saturated calomel (SCE): using KCl(satd): sodium saturated calomel (SSCE): using NaCl(satd).
[b]Standard hydrogen electrode (SHE): $a_{H^+} = 1$, $p_{H2} = 1$ atm.

Reference electrodes have large concentrations of the ions and large reservoirs of the solid, liquid and gaseous reagents necessary for the redox reaction. Small changes in the amounts of these species make little difference to their activities and make little difference to the electrode potential (see Topic E5). The standard hydrogen electrode (SHE) is not a very practical reference electrode, as the platinum electrode is readily poisoned, hydrogen gas is explosive and requires bulky cylinders and the electrode potential is sensitive to small changes in gas pressure (see Topic E5), so alternative reference electrodes are used for experimental measurements.

**Standard reduction potentials**

Variations in the potential of different half-cells can be used to compare the propensity for oxidation or reduction in the half-cell reactions. To do this, a **standard reduction potential** or **standard potential** is measured for a half-cell. This is the standard cell potential of a cell consisting of the half-cell as the right-hand electrode and the SHE as the left-hand electrode, separated by a salt bridge. These are denoted as $E^\ominus(ox1, ox2,../red1, red2..)$ where ox1, ox2.. are the oxidized species and red1, red2,.. the reduced species in the half-cell reaction. Values are typically measured and tabulated at 25°C or 298 K. The **standard cell potential** for any cell can then be calculated, as it is simply the difference between the RH and LH half-cell standard potentials:

$$Pt \mid Fe^{2+}(aq), Fe^{3+}(aq) \parallel Zn^{2+}(aq) \mid Zn(s)$$

$$E^\ominus_{cell} = E^\ominus(Zn^{2+}/Zn) - E^\ominus(Fe^{3+}/Fe^{2+})$$

*The electrochemical series*
When the two electrodes are connected and current is allowed to flow in a galvanic cell, the formal cell reaction as written is **spontaneous** when $\Delta G$ is negative (see Topic B6), which is when reduction occurs at the RH electrode and oxidation at the LH electrode. This happens when the RH electrode potential is more positive than the LH electrode, or when $E^\ominus_{cell} > 0$. The reverse reaction is

spontaneous when $E^{\ominus}_{cell} < 0$ and when $E^{\ominus}_{cell} = 0$, the cell is at equilibrium. This means that the reduced form of a couple with a low $E^{\ominus}$ value will reduce the oxidized form of a couple with a higher $E^{\ominus}$ value. For $E^{\ominus}(F_2/F^-) = + 2.87$ V and $E^{\ominus}(Br_2/Br^-) = + 1.09$ V, the spontaneous reaction will therefore be:

$$F_{2(g)} + 2Br^-_{(aq)} \rightarrow 2F^-_{(aq)} + Br_{2(l)}$$

and values of $E^{\ominus}$ when tabulated in order give an **electrochemical series** which shows an increase in the oxidizing power of the oxidizing agent (and a corresponding decrease in the reducing power of the reducing agent) in the redox couple as $E^{\ominus}$ increases.

**Thermodynamic data**

The **change in Gibbs free energy** (see Topic B6) is given by the energy change of the electrons travelling across the cell voltage:

$$\Delta G_{cell} = -nFE_{cell}$$

where $n$ is the number of moles of electrons transferred in the cell reaction (equal to the number of electrons in each half-cell reaction when the cell reaction is obtained) and $F$ is **Faraday's constant**, the charge on a mole of electrons.

For this and all other thermodynamic equations in this section, under the particular condition of standard states this general equation applies. This

$$\Delta G^{\ominus}_{cell} = -nFE^{\ominus}_{cell}$$

and since (see Topic C1),

$$\Delta G^{\ominus}_{cell} = -RT\ln K$$

measurement of $E^{\ominus}_{cell}$ allows calculation of K, the equilibrium constant for the **formal cell reaction** (see Topic E5).

The **change in entropy**, $\Delta S_{cell}$, due to the cell reaction at any temperature is given by:

$$\Delta S_{cell} = -\left(\frac{\partial \Delta G_{cell}}{\partial T}\right)_p = nF\left(\frac{\partial E_{cell}}{\partial T}\right)_p$$

and the **change in enthalpy**, $\Delta H_{cell}$, is:

$$\Delta H_{cell} = \Delta G_{cell} + T\Delta S_{cell} = -nFE_{cell} + nFT\left(\frac{\partial E_{cell}}{\partial T}\right)_p$$

The **change in volume**, $\Delta V_{cell}$, for the cell reaction is:

$$\Delta V_{cell} = \left(\frac{\partial \Delta G_{cell}}{\partial p}\right)_T = -nF\left(\frac{\partial E_{cell}}{\partial p}\right)_T$$

These can be calculated from measurements of $E_{cell}$ and its variation with temperature, $T$, at constant pressure, $p$, or with pressure at constant temperature around the conditions of interest.

These relationships also apply to the standard half-cell reactions, as $\Delta G^{\ominus}$, $\Delta H^{\ominus}$ and $\Delta S^{\ominus}$ are zero for the SHE half-cell reaction $H^+_{(aq)} + e^- \rightarrow \frac{1}{2}H_{2(g)}$ by definition, so that:

$$\Delta G^{\ominus} = -nFE^{\ominus},$$

$$\Delta S^{\ominus} = nF\left(\frac{\partial E^{\ominus}}{\partial T}\right)_p,$$

$$\Delta H^\ominus = -nFE^\ominus + nFT\left(\frac{\partial E^\ominus}{\partial T}\right)_p \text{ and}$$

$$\Delta V^\ominus = -nF\left(\frac{\partial E^\ominus}{\partial p}\right)_T$$

are the equations for calculating the change in standard Gibbs free energy, entropy, enthalpy and volume of the half-cell reduction reaction. This allows $\Delta G^\ominus$, $\Delta H^\ominus$ and $\Delta S^\ominus$ values to be calculated from the temperature and pressure variation of the standard reduction potentials. $\Delta V^\ominus$ values are small for half-cells involving solid, liquid and ionic reagents, which means that half-cell potentials are relatively insensitive to pressure changes. However, volume changes are large for gas electrodes such as the hydrogen electrode, which is the origin of the sensitivity of its potential to small changes in pressure.

# E5 ELECTROCHEMISTRY AND ION CONCENTRATION

## Key Notes

**Activity dependence of cell voltage**
The cell voltage and half-cell potentials depend on the activity of the ions in the cell reaction and the half-cell reaction. The relationships linking cell voltage and activity and half-cell potential and activity are the Nernst equations for cell and half-cell respectively.

**Cells at equilibrium**
An electrochemical cell at equilibrium has no cell voltage. This means that the equilibrium constant for the cell reaction can be obtained from the standard cell voltage, $E_{cell}^{\ominus}$.

**Measurement of activity coefficients**
Measurements of cell voltages as a function of ionic concentration can be used to determine the mean activity coefficient of the ions in the cell reaction at each concentration. The standard cell voltage can also be found, as the ionic concentrations tend to zero.

**Electrolyte concentration cells**
Electrolyte concentration cells have a cell reaction that involves the movement of an ion from a solution of one concentration to another. The Nernst equation for this cell allows the free energy change for this process to be calculated. The cell voltage corresponds to the voltage set up between these two solutions across a membrane which allows only this ion to pass through it.

**pH dependence of cell voltage**
Cells that contain the hydrogen electrode half-cell show a dependence of the cell voltage on pH given by the Nernst equation. This equation can be used to convert between the standard state and the biological standard state. A cell consisting of the hydrogen electrode and a reference electrode could in principle be used to measure changes in the pH of a solution. Practical difficulties mean that a cell consisting of the glass electrode and a reference electrode is preferred.

**Related topics**
Thermochemistry (B3)
Free energy (B6)
Fundamentals of equilibria (C1)

Thermodynamics of ions in solution (E2)
Electrochemical thermodynamics (E4)

**Activity dependence of cell voltage**
Measurement of cell potentials away from the standard state provides a convenient method for measuring ion **activity** and **activity coefficients** (see Topic E2). As ion concentration is varied in a half-cell to vary ion activity, the equilibrium position of the half-cell reduction reaction changes, and this causes a change in the half-cell potential. The relationship between the half-cell potential and activity is given by the **Nernst equation**:

$$E = E^{\ominus} - \frac{RT}{nF} \ln\left(\frac{a_C^c a_D^d}{a_A^a a_B^b}\right)$$

for the general half-cell reduction reaction:

$$aA + bB + ne^- \rightarrow cC + dD$$

where $a_A^a$, $a_B^b$, $a_C^c$, $a_D^d$ are the activities of A, B, C, D raised to the power of their stoichiometries and $n$ is the number of electrons transferred (see Topic E4). For the formal cell reaction, $aA + bB \rightarrow cC + dD$, the Nernst equation becomes

$$E_{cell} = E_{cell}^{\ominus} - \frac{RT}{nF} \ln Q \qquad Q = \left(\frac{a_C^c a_D^d}{a_A^a a_B^b}\right)$$

where $Q$ is the **reaction quotient** for the cell reaction in terms of their activities (see Topic B6) and $n$ is the number of electrons in each half-cell reaction used to obtain the **formal cell reaction** (see Topic E4). The activity of pure liquids and solids is equal to 1, which simplifies both Nernst expressions (see Topic C1).

**Cells at equilibrium**

When a cell is at equilibrium, $E_{cell} = 0$ (the **'flat battery'** condition) and $Q = K$, the **equilibrium constant** for the reaction (see Topic C1). This gives the expression:

$$E_{cell}^{\ominus} = \frac{RT}{nF} \ln K$$

which allows calculation of equilibrium constants (see Topic C1) for cell reactions from calculated or measured $E_{cell}^{\ominus}$ values (see Topic E4).

**Measurement of activity coefficients**

For a cell such as M(s) | M⁺(aq, c) || Cl⁻ (aq, c) | AgCl(s) | Ag (s) , the cell reaction is $AgCl_{(s)} + M_{(s)} \rightarrow Ag_{(s)} + M^+_{(aq)} + Cl^-_{(aq)}$ and the Nernst equation for the cell is:

$$E_{cell} = E_{cell}^{\ominus} - \frac{RT}{F} \ln\left(a_{M^+} a_{Cl^-}\right)$$

From the relationship of activity to concentration (see Topic E2):

$$E_{cell} = E_{cell}^{\ominus} - \frac{2RT}{F} \ln\left(\frac{c}{c^{\ominus}}\right) - \frac{2RT}{F} \ln(\gamma_{\pm})$$

where $c^{\ominus}$ is the standard concentration of 1 mol dm⁻³. Measurements of $E_{cell}$ as the concentration $c$ is varied therefore allow measurements of $\gamma_{\pm}$ (see Topic E2) if $E_{cell}^{\ominus}$ is known or calculated. $E_{cell}^{\ominus}$ can be determined from tabulated data (see Topic E4) or by determining $E_{cell}^{\ominus}$ as $c \rightarrow 0$ since $\gamma_{\pm} \rightarrow 1$ (see Topic E2).

**Electrolyte concentration cells**

An **electrolyte concentration cell** involves connecting two identical half-cells via a **salt bridge**, for example M(s) | M⁺ (aq, a = x) || M⁺ (aq, a = y) | M(s). The only difference in the two half-cells is the difference in ion activity, $x$ and $y$. The cell reaction is M⁺ (aq, a = y) → M⁺ (aq, a = x) and the Nernst equation is:

$$E_{cell} = -\frac{RT}{F} \ln\left(\frac{x}{y}\right)$$

as $E^{\ominus}$ is the same for both half-cells. This reaction also occurs when there is a membrane between the two M⁺ solutions across which only M⁺ can transfer. An example is the junction between the inside and outside of a neuron cell membrane, where an imbalance in K⁺ activity is developed, producing the potential difference $E_{cell}$ that drives the nerve impulse. The potential is positive in the solution of high M⁺ concentration and negative in the low M⁺ solution and is established to increase the rate of transfer from high concentration to low and equalize the M⁺ activity.

**pH dependence of cell voltage**

Cells that incorporate the **hydrogen gas electrode** have a cell potential which is dependent on pH. For example Pt | H$_2$(g, p = 1 atm) | H⁺(aq) ‖ Cl⁻ (aq) | AgCl(s) | Ag(s) has a cell reaction AgCl$_{(s)}$ + ½H$_{2(s)}$ → Ag$_{(s)}$ + H⁺$_{(aq)}$ + Cl⁻$_{(aq)}$ and the cell Nernst equation is:

$$E_{cell} = E_{cell}^{\ominus} - \frac{RT}{F}\ln\left(a_{H^+} a_{Cl^-}\right)$$

$$= E_{cell}^{\ominus} - \frac{RT}{F}\ln a_{Cl^-} - \frac{2.303RT}{F}\log_{10} a_{H^+}$$

$$= E_{cell}^{\ominus} - \frac{RT}{F}\ln a_{Cl^-} + \frac{2.303RT}{F}pH$$

(see Topic C2). The cell Nernst equation can then be used to calculate the difference between the cell potential in the standard state, when pH = 0, and the biological standard state, when pH = 7 (see Topic B2). The RH electrode has a constant concentration of Cl⁻ and is a reference electrode (see Topic E4); the first two terms on the right-hand side of the equation are therefore constant and the cell potential could be used to measure changes in the pH of the solution in the LH half-cell. In fact, using the hydrogen gas electrode to measure pH is impractical (for the same reasons as the **standard hydrogen electrode** (**SHE**) is impractical, see Topic E4). Instead, solution pH measurements are made using the **glass electrode** (*Fig. 1*).

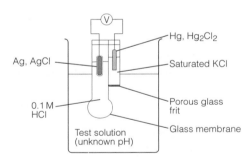

*Fig. 1.    The glass electrode cell for measuring pH.*

This consists of a glass membrane between two reference electrodes:

Ag(s) | AgCl(s) | HCl(aq, 0.1 M) | glass | H⁺(aq, pH = ?) ⦙ Cl⁻(aq, sat) | Hg$_2$Cl$_2$(s) | Hg(l)
test solution

This cell is combined into a single probe, as shown in *Fig. 1*, and dipping it into the test solution of H⁺ of unknown pH completes the circuit. Both LH and RH electrodes are reference electrodes (the **silver–silver chloride** and **saturated calomel electrode** (**SCE**), see Topic E4) and maintain a constant potential, as does the **liquid junction potential** due to the glass frit. The cell voltage therefore changes only in response to the changes in the glass membrane voltage. The glass membrane is impermeable to protons, but has sites that exclusively take up protons from solution, changing the cation distribution in the membrane.

Potential differences at the two membrane–solution boundaries are set up as a result of this process. The size of these voltages depends on the pH in solution, as more $H^+$ will produce more proton uptake and a higher voltage. A constant pH on the left-hand side of the membrane ensures a constant voltage here, but changes in pH in the test solution cause a corresponding voltage change. The cell Nernst equation is:

$$E_{cell} = E' + \frac{2.303RT}{F}\text{pH}$$

For any glass electrode, the validity of this expression is determined and $E'$, the constant in the Nernst equation, is found by **calibration**, which involves measuring $E_{cell}$ for two (or more) **buffer solutions** of known pH (see Topic C4). Once calibrated an $E_{cell}$ measurement can be used to determine the pH of any unknown solution.

# E6 MACROSCOPIC ASPECTS OF IONIC MOTION

## Key Notes

**Conductance and conductivity**

The conductance of an ionic solution is measured by applying an oscillating voltage between two parallel plate electrodes, which avoids concentration polarization and ensures the current is proportional to the applied voltage. The conductance is the ratio of the current to the voltage. The conductance, $L$, is used to calculate the conductivity, $\kappa$, which is a measure of the charge carrying ability of the ionic solution and is independent of the cell in which the measurement is performed.

**Molar conductivity**

The conductivity is proportional to the number of ions in solution. The molar conductivity is $\kappa/c$, where $c$ is the salt concentration, and gives a measure of the charge carrying capabilities for the same amount of dissolved salt. This allows comparison of the charge carrying capabilities for the same amount of different dissolved electrolytes. The molar conductivity is an addition of the molar conductivities of the cations and anions.

**Strong electrolytes**

A strong electrolyte is an electrolyte that completely dissociates into its constituent ions. Its molar conductivity slowly decreases with increasing concentration, due to the increasing importance of ionic interactions.

**Limiting molar conductivity**

The limiting molar conductivity of an electrolyte is the molar conductivity as $c \to 0$, where there are no ionic interactions. The limiting molar conductivity is a combination of the limiting molar conductivities of the cations and of the anions. The limiting molar conductivity of a particular ion is constant.

**Weak electrolytes**

Weak electrolytes do not completely separate into their constituent ions except at high dilution. As $c$ increases, the molar conductivity falls relatively rapidly, as the proportion of undissociated electrolyte increases. Molar conductivity measurements allow the degree of electrolyte dissociation to be calculated at any $c$.

**Transport numbers**

The transport number of an ion is a measure of the fraction of the total current carried by the ion. The sum of the transport numbers for the ions in solution add up to 1. The transport number for an ion varies with the nature of the counterion(s) and with $c$.

**Related topics**

Solubility (C5)
Ions in aqueous solution (E1)

Thermodynamics of ions in solution (E2)

**Conductance and conductivity**

When an electric field is put across an aqueous ionic solution by applying a voltage between two parallel plate electrodes, the cations are attracted towards the negative plate (cathode) and the anions towards the positive plate (anode).

This movement of ions in a field is called **migration** and the movement of charge results in a current in the electric circuit connecting the electrodes. The field between the plates is initially linear and produces a constant ion velocity at all points. However, if there is no redox reaction, cations and anions will separate and collect at the cathode and anode respectively with time, balancing the electrode charge and there will be fewer ions in the bulk of solution. This is a process known as **concentration polarization**, which also leads to modification of the original linear potential profile between the two electrodes and a variation of ion velocity between the plates. The observed current will also fall with time (as at infinite time, the ion flow will have stopped as the ions will have reached the plates and the current will have decreased to zero). Concentration polarization is therefore useful when separating ions is desirable, such as in **electrophoresis** and **electro-osmosis**, but this is to be avoided when measuring fundamental ion motion.

For these measurements, the field polarity is rapidly switched (at around 1000 times per second) by applying an AC (alternating) voltage to the plates (*Fig. 1*).

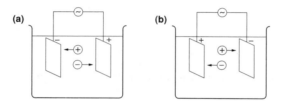

*Fig. 1.    The effect of an AC voltage on the ion motion as electrode polarity is switched.*

The ions then alternately migrate first to one plate and then the other during cycling, so that the ions oscillate around a fixed position, avoiding concentration polarization.

Under these conditions the observed current, $i$, is always proportional to the applied voltage, $V$, with $i = VL$ and the constant of proportionality being the **conductance**, $L$. $L$ depends on the experimental apparatus; the area, $A$, of the plate electrodes and the distance, $l$, between them and for parallel field lines, with two plates exactly opposite each other $\kappa = L(l/A)$, with $\kappa$ being the **conductivity**. It is the conductivity that probes the rate of charge transfer in the solution free from variations in apparatus design. In practice it is very difficult to construct experimental conductivity cells which exactly obey this equation, but generally $\kappa = LC$, where $C$ is the **cell constant** which can be determined by calibration. This involves determining $L$ for a solution of known $\kappa$, calculating $C$ and using it to determine $\kappa$ from $L$ for all other solutions.

**Molar conductivity**

The rate of charge transfer is proportional to the number of ions in the solution between the plates, which itself will vary with the concentration, $c$, of dissolved electrolyte. The **molar conductivity**, $\Lambda_m$, is defined as:

$$\Lambda_m = \frac{\kappa}{c}$$

and allows for this variation with concentration. $\Lambda_m$ values enable comparison of the conductivities (or charge carrying capabilities) of an equivalent number of moles of electrolyte both as the type and the concentration of electrolyte is varied. These measurements allow variation in ion migrational motion in these systems to be studied.

**Strong electrolytes**

A **strong electrolyte** completely dissociates into its constituent ions. An example is NaCl. It might be expected that $\Lambda_m$ would be independent of concentration for a strong electrolyte, but in reality for dilute solutions

$$\Lambda_m = \Lambda_m^0 - \kappa\sqrt{c}$$

where $\kappa$ is a constant for a particular electrolyte and $c$ is the electrolyte concentration (*Fig. 2a*).

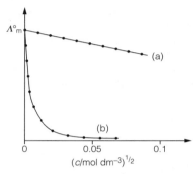

Fig. 2.    The dependence of molar conductivity on concentration for (a) a strong electrolyte; (b) a weak electrolyte.

As $c$ increases, the cations and anions migrate more slowly through solution. This is because the anions and cations, moving in opposite directions, are closer together and their electrostatic attraction grows in importance, which progressively slows ion progress. This is a consequence of the development of an ionic atmosphere around the ions, which explains the dependence of molar conductivity on $\sqrt{I}$ (where $I$ is the **ionic strength**) and hence $\sqrt{c}$ (see Topics E1 and E2).

Charge is carried in the solution both by cations moving towards the cathode and anions moving in the opposite direction towards the anode, and so the molar conductivity is simply a combination of the molar conductivities of the cation, $\lambda_+$, and of the anion, $\lambda_-$:

$$\Lambda_m = n_+\lambda_+ + n_-\lambda_-$$

where $n_+$ and $n_-$ are the number of moles of cations and anions per mole of salt, i.e. for an electrolyte with the overall molar formula $M_{n+}X_{n-}$. The degree of interaction between cation and anion depends on the charge and size of the ions (see Topic E2) and so $\lambda$ for an ion often varies as the counterion is varied.

**Limiting molar conductivity**

For non-interacting ions, i.e. when $c \to 0$ and the cations and anions are so far apart in solution they do not interact with each other, the molar conductivity is called the **limiting molar conductivity**, $\Lambda_m^0$ (*Fig. 2*). This is a combination of the limiting molar conductivities of the cation, $\lambda_+^0$, and the anion, $\lambda_-^0$:

$$\Lambda_m^0 = n_+\lambda_+^0 + n_-\lambda_-^0$$

and under these conditions, $\lambda^0$ values are constant for any ion and do not vary when the counterion is varied. They have been tabulated for many ions, which allows values of $\Lambda_m^0$ to be calculated for different salts.

**Weak electrolytes**

For a **weak electrolyte**, the salt does not completely dissolve into its constituent ions and for example the equilibrium:

$$MA_{(aq)} \overset{K}{\rightleftharpoons} M^+_{(aq)} + X^-_{(aq)}$$

is present. Dilute concentrations favor the formation of ions and as $c \rightarrow 0$, $\Lambda_m \rightarrow \Lambda_m^0$ because complete dissociation occurs. Increasing the concentration decreases the fraction of dissociated ions (the **degree of dissociation, $\alpha$**) in accordance with **Le Chatelier's principle** (see Topic C1). Undissociated electrolyte is uncharged and cannot contribute to the conductivity and so the molar conductivity falls much more rapidly with concentration than for a strong electrolyte (*Fig. 2b*). As this fall is much larger than that due to the electrostatic attraction of ions, the fraction of salt present as ions, $\alpha$, at any electrolyte concentration $c$ is given by $\alpha \approx \Lambda_m/\Lambda_m^0$.

$\Lambda_m$ is measured at $c$, and in the example if $\Lambda_m^0$ is measured or calculated, $\alpha$ can be used to determine the concentration of cation and anion (each $\alpha c$) and undissociated electrolyte ($c - \alpha c$) and obtain a value for the equilibrium constant, $K$, for the ion dissociation reaction (see Topic C5). In the example, at low ionic strength (see Topic E2),

$$K \approx \frac{c_{M^+}c_{X^-}}{c_{MX}c^{\ominus}} = \frac{\alpha^2 c}{(1-\alpha)c^{\ominus}}.$$

**Transport numbers**

The **transport number, $t$**, of an ion is the fraction of the total charge carried by the ion in an electrolyte. For the cation, and the anion, these are given by

$$t_+ = n_+\lambda_+/\Lambda_m \text{ and } t_- = n_-\lambda_-/\Lambda_m$$

respectively, and the sum of the transport numbers for the cations and the anions must add up to 1. Since $\lambda$ and $\Lambda_m$ vary with concentration and electrolyte, so does $t$. The **limiting transport number, $t^0$**, is the value measured when $c \rightarrow 0$.

# E7 MOLECULAR ASPECTS OF IONIC MOTION

---

## Key Notes

**Ion mobility**

The mobility of an ion is its terminal speed in a unit applied field. The mobility can be calculated from the ionic molar conductivity and also varies with concentration. The limiting mobility as $c \rightarrow 0$ can be used to determine the hydrodynamic radius of the ion.

**Hydrodynamic radius**

The hydrodynamic radius is the radius of the ion as it moves through solution with its solvation shell. This radius is significantly larger than the ion radius in the gas phase. The solvation shell size can dominate the hydrodynamic radius for ions which are small and highly charged when unsolvated. This can lead to a larger hydrodynamic radius for these ions compared with intrinsically larger or less charged ions.

**H⁺/OH⁻ mobility**

$H^+$ and $OH^-$ have anomalously high mobilities for ions of their size in water. This is because movement of $H^+$ and $OH^-$ ions through solution is achieved by bond rearrangement in the hydrogen-bonded water framework of the solvent. This is a much faster process than the physical movement of ions through the solution.

**Related topic**

Ions in aqueous solution (E1)

---

**Ion mobility**

**Ion mobility** is related to molar conductivity by the equations:

$$\lambda_+^0 = u_+^0 z_+ F, \quad \lambda_-^0 = u_-^0 |z_- F|,$$
$$\lambda_+ = u_+ z_+ F \quad \text{and} \quad \lambda_- = u_- |z_- F|$$

which allows the **mobilities, $u_+$** and **$u_-$** of the cation and anion to be determined from **molar conductivity** measurements at and away from infinite dilution. $z_+$ and $z_-$ are the **formal charges** of the cation and anion respectively, so $z_+ F$ and $|z_- F|$ are the magnitudes of the charges on a mole of cations and anions. The mobility is always positive and is a measure of the **terminal migration speed** of an ion per unit applied field. This limiting speed is attained when the acceleration due to the field is exactly balanced by the viscous drag of the ion moving through the solution, which for a spherical ion leads to the equations:

$$u_+^0 = \frac{z_+ e}{6\pi\eta a} \quad \text{and} \quad u_-^0 = \frac{|z_- e|}{6\pi\eta a}$$

where $e$ is the charge on the electron, so $ze$ is the charge on the ion, $\eta$ is the **viscosity**, a constant for any solvent which determines how easy it is for the ion to part the solvent molecules and move through solution and $a$ is the **hydrodynamic radius** of the solvated ion.

**Hydrodynamic radius**

The **hydrodynamic radius** is the radius of the ion as it migrates through solution. At the low applied fields typically used for conductivity measurements, this closely mirrors the radius of the solvated ion (see Topic E1), as the forces between ion and solvent in the **solvation shell** are sufficiently strong to ensure the ion moves with its solvation shell. This means that the hydrodynamic radius is typically much larger than the ion radius in the gas phase. The smallest, most highly charged ions (such as $Li^+$, $Al^{3+}$ and $F^-$) before solvation have the largest solvation shells (see Topic E1). Since the overall radius of the solvated ion is the sum of the ionic radius plus the solvation shell radius, the smallest unsolvated cations often have the largest radii when solvated and move slowest through solution. Singly-charged anions often have similar hydrodynamic radii, as they tend to be larger than singly-charged cations, have smaller solvation shells, and the effect of a change in the size of the ion is often counterbalanced by the change in the solvation shell radius.

**H⁺/OH⁻ mobility**

Protons and hydroxide ions have anomalously high ionic molar conductivities and mobilities in comparison to all other ions, and in particular for their size. This is as a result of the mechanism by which they move through solution, often called the **Grotthus mechanism** (*Fig. 1*).

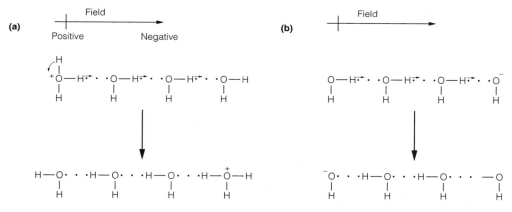

*Fig. 1.* The Grotthus mechanism for (a) H⁺; (b) OH⁻ ion motion in water. The arrows indicate the concerted proton movement when the field is applied.

Instead of the movement of a proton (present in water as the $H_3O^+$ species) or the OH⁻ ion physically through solution, a concerted field-driven bond redistribution in the hydrogen-bonded water network results in the destruction of the ion and its production in another location. Thus, ion movement is achieved by an electronic redistribution with little nuclear redistribution, which much enhances the transfer rate compared with other ions.

# E8 THE MOTION OF LARGER IONS

---

## Key Notes

| | |
|---|---|
| **The electrical double-layer** | Large ions are often multiply charged and are surrounded by an electrical double-layer, consisting of a Stern layer of tightly bound counterions and a diffuse double-layer which is an ionic atmosphere containing a majority of counterions. As with smaller ions, this ionic atmosphere is crucial for determining ionic interactions. The residual charge at the outside edge of the Stern layer is the effective charge on the ion, which can vary with electrolyte concentration and type. |
| **Electrophoresis** | Electrophoresis is the separation of large ions by applying a field between two plates to induce concentration polarization. Under the influence of the field, ions that have small hydrodynamic radii and large effective charges move further and separate from larger, less charged ions. |
| **Electro-osmosis** | Electro-osmosis is a method for inducing osmosis by an electric field. It is the counterpart to electrophoresis and involves applying a field across a solution with fixed larger ions. Motion of the counterions is induced, which also produces water flow (osmosis). As with electrophoresis, the rate of water flow is controlled by the effective charge on the ion. |
| **Related topics** | Ions in aqueous solution (E1)    Macroscopic aspects of ionic motion (E6) |

---

**The electrical double-layer**

Larger ions, such as biological polymers or macromolecules, which can contain many thousands of atoms and many ionizable groups, are often multiply charged. The resulting electrostatic forces between these ions and their small counterions are often sufficiently strong that an **electrical double-layer** is formed (*Fig. 1*).

Near the large ion, a layer of counterions (often partially desolvated) is held tightly to the surface by the very strong attractive electrostatic forces. This is termed the **Stern layer**, and these strongly adsorbed counterions reduce the effective charge of the ion. Outside the Stern layer, the remaining charge imbalance is sufficiently small that an **ionic atmosphere** (see Topic E1) is formed around the ion called a **diffuse double-layer**, which contains an overall excess of counterions sufficient to balance the residual charge, loosely bound to the ion by the electrostatic forces of attraction. As with smaller ions, the interaction between the ion and its ionic atmosphere of diffuse double-layer is crucial in determining the ion thermodynamics and ionic motion. This is governed, not by the charge on the ion, but by the effective charge on the ion at the outside edge of its Stern layer. Thus, the use of strongly adsorbing counterions or an increase in counterion concentration, each of which have a tendency to give a larger number of ions in the Stern layer, leads to a decrease in the effective ion charge.

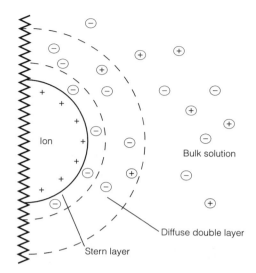

*Fig. 1.*   *The structure of the electrical double layer around a large ion.*

**Electrophoresis**

**Electrophoresis** is a method for separating large ions of different charge and size, and involves applying a constant voltage between the plates to utilize the concept of **concentration polarization** (see Topic E6). In **zone electrophoresis**, a gel or solid support (such as a piece of filter paper) is used to minimize the movement of solvent and a constant field is applied between two plate electrodes. In order to aid separation, generally the mixture of large ions is applied as a localized band or spot onto the solid support. On application of the field, smaller, more highly charged ions move more rapidly and in time separate from larger, less highly charged ions and produce separate, identifiable bands or spots on the electrophoresis support. As all these large ions have low mobility, the field is generally large in order to induce appreciable ion separation. This field is usually sufficient to overcome the electrostatic force between the ion and its diffuse double-layer and cause **shearing** (or movement in different directions) at the outside of the **Stern layer**. The large ion with its Stern layer therefore moves towards one electrode at a rate governed by its effective charge and its solvated ion size and the diffuse double-layer (of opposite charge) moves towards the other electrode. The mobility of the ion, complete with its Stern layer, can therefore be varied by judicious choice of counterion type and concentration.

For further separation of large ions such as biomolecules that have similar size and charge, a two-stage electrophoresis method is often used (*Fig. 2*). This takes advantage of the fact that these large molecules often have many ionizable groups such as $R-COOH \rightleftharpoons R-COO^- + H^+$ and $R-NH_3^+ \rightleftharpoons R-NH_2 + H^+$ where R is the biomolecule. This means that the position of these equilibria, and hence the degree of ionization and the charge on the ion, depend upon the solution pH. If, after separation by electrophoresis, two or more different biomolecules with similar size and charge have not separated and remain together in a particular spot, a second electrophoresis experiment is then carried out. This involves applying the field to the separated spots at an orientation of 90° to the first field in a solution of different pH. The biomolecules within the same spot, each of which contains its own characteristic number and

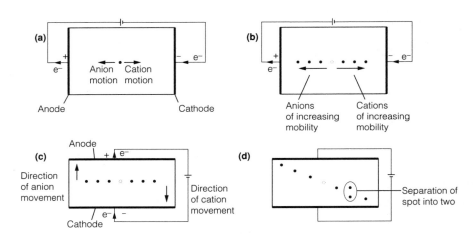

*Fig. 2. The two-stage electrophoresis method. (a) Initial spot application; (b) after first electrophoresis separation; (c) application of second electrophoresis field; (d) after second electrophoresis separation.*

distribution of ionizable groups, will now have a different effective charge and separation of the spot into separate spots, each containing an individual biomolecule, will occur.

**Electro-osmosis**   **Electro-osmosis** is a complementary technique to **electrophoresis** and is a way of inducing water flow (**osmosis**) by the application of an electric field. Large ions are trapped in a gel or matrix so that they cannot move (or the gel or matrix simply has a charged surface). Application of a large constant voltage between two electrodes either side of the gel results in the motion of the diffuse double-layers of counterions towards the plate electrode of opposite charge, as with electrophoresis. These counterions move with their associated solvent and induce a net flow of water across the gel or matrix. This is termed **electro-osmosis** (electrically induced osmosis) and the rate of water flow is determined by the effective charge on the large ion which, as with electrophoresis, can be controlled by varying counterion type, concentration and electric field.

# F1 EMPIRICAL APPROACHES TO KINETICS

## Key Notes

**Experimental methods**

Experimental methods in kinetics measure change in the composition of a reaction mixture with time, either continuously as the reaction progresses, or at fixed intervals after the reactants have come together. The techniques applied vary depending on the timescale of the reaction and the chemical species under study. Additional kinetic information is obtained by varying experimental parameters such as the initial concentration of reactant(s) or the temperature of the mixture.

**Rate of reaction**

The instantaneous rate of reaction for a species is the rate of change of concentration with time of that species at a particular instant during the reaction. Units of reaction rate always have dimensions of concentration time$^{-1}$.

**Rate law**

The rate law is the empirically determined mathematical relationship describing the observed rate of reaction in terms of the concentrations of the species involved in the reaction. Rate laws do not necessarily fit the simple stoichiometry of a balanced chemical reaction but may be the consequence of a more complex underlying mechanism to the observed reaction.

**Rate constants**

Rate constants are the constants of proportionality within the empirical rate law linking rate of reaction and concentration of species involved in the reaction. The units of rate constants are particular to the rate law and can be derived by dimensional analysis. Rate constants usually vary with temperature.

**Order of reaction**

If the rate law for a reaction can be written in the form, rate $\propto [A]^{\alpha}[B]^{\beta}$... then the reaction is classified as $\alpha$-order in A, $\beta$-order in B,... and as $(\alpha+\beta+...)$-order overall. The exponents do not have to be integers, and for complex rate laws, the order may not be a definable quantity.

**Molecularity**

The molecularity of a reaction is the number of molecules which come together to react and is independent of the order of a reaction. In a unimolecular reaction a single molecule breaks apart or rearranges its constituent atoms. A bimolecular reaction involves two atoms or molecules.

**Related topics**

**Experimental methods**

Chemical kinetics is the study of the **rate** at which chemical reactions occur. Although the timescales of chemical reactions vary enormously, ranging from days or years to just a few femtoseconds ($10^{-15}$ s), the basic principle of all experimental kinetic methods is the same. Reactants of particular concentration are brought together and some measure is made of the rate at which the composition changes as the reaction progresses. Depending on the specificity of the detection method available, monitoring the rate of reaction may involve measurement of the rate at which specific reactants (or a subset of reactants) are consumed and/or the rate at which specific products (or a subset of products) are formed, or simply measurement of some bulk property of the system such as pressure, pH or ionic conductivity. More sophisticated detection might involve chromatography, mass spectrometry, or optical techniques such as absorption, fluorescence or polarimetry.

In a real-time method the composition of the system is analyzed while the reaction is in progress, either by direct observation on the mixture or by withdrawing a small sample and analyzing it. In the latter case it is important that analysis is fast compared with the rate of continuing reaction. Alternatively the composition of the bulk (or of a sample withdrawn from it) may be analyzed after the reaction has been deliberately stopped or **quenched**. Quenching might be achieved by diluting the mixture rapidly in excess solvent, neutralizing with acid or sudden cooling. Quenching is only suitable for reactions which are slow enough for there to be little reaction during the time it takes to quench the mixture.

In the **continuous flow** method reactants are mixed together at a single point as they flow through a reaction vessel. The point of mixing establishes time zero and the reaction continues as the reagents continue to flow down the tube (*Fig. 1*). The distance downstream at which analysis occurs defines the time

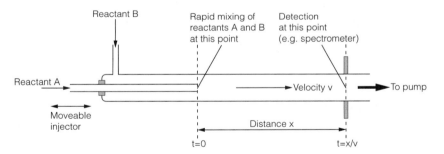

*Fig. 1.   Schematic of apparatus for measurement of reaction rates in the gas-phase by the method of continuous flow. The technique is similar for continuous flow measurements of reactions in liquid.*

since reaction initiation and this can be varied either by moving the point of analysis or the point of mixing of the reagents. A disadvantage of continuous flow is that it uses a large amount of reagent. The **stopped-flow** method overcomes this by injecting the reagents very quickly into a reaction chamber designed to ensure rapid mixing. Beyond the reaction chamber there is an observation cell and a plunger which moves backwards to accommodate the incoming mixture and stops the flow when a pre-determined volume has been filled (*Fig. 2*). The filling of the chamber corresponds to the initial preparation of mixed reagents and the course of the reaction is monitored in the observation

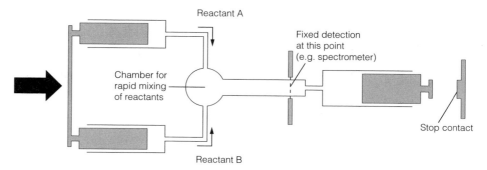

*Fig. 2. Schematic of apparatus for measurement of reaction rates in the liquid phase by stopped flow.*

cell. Since only a single fill of the chamber is prepared the method uses less material than the continuous flow method. It is widely used to study **enzyme kinetics** (see Topic F6).

Reactions with duration of seconds or less, for example reactions involving radicals, are often investigated using the technique of **flash photolysis**. A precursor mixture is exposed to a brief flash of light to generate an initial concentration of one of the reactants by photolysis and the contents of the chamber are then monitored spectroscopically, either continuously, or at discrete time intervals after the flash. The use of **lasers** and electronic acquisition of data enables study of reactions faster than a few tens of microseconds or less, and with a high degree of reagent or product specificity.

Interpretation of experimental data is often simplified by the **method of isolation** in which the concentration of one reagent is kept constant whilst the concentration of the other reagent(s) are varied in turn. An extension of this approach is to ensure that the initial concentration of other reagents are in large excess of the one being monitored so that the concentrations of the former remain effectively constant during the reaction. This gives rise to **pseudo rate laws** (see Topic F2).

Whatever experimental method is used the reaction must be maintained at a constant temperature throughout otherwise the observed rate is a meaningless amalgamation of the different rates at different temperatures. However, systematically repeating the experiment at different temperatures provides additional information on the **activation energy** and **Arrhenius equation** for the reaction (see Topic F3).

**Rate of reaction**

The **rate of reaction** of a designated species is the rate of change of concentration of that species with time. Since rates of reaction usually vary during a reaction, because of changing concentrations of reagents, it is necessary to consider **instantaneous rates of reaction** evaluated at specific instants during the reaction (e.g. the **initial rate** of reaction when the reagents are first mixed). Rate of reaction is therefore equal to the gradient of the curve of species concentration against time evaluated at the time of interest. The steeper the gradient, the greater the rate of reaction (*Fig. 3*). A species which is being consumed has a negative gradient, whilst a species which is being formed has a positive gradient. The units of rate of reaction always have dimension of concentration time$^{-1}$.

Reactants and products may be consumed and formed at different rates according to the particular reaction **stoichiometry** (the numbers of molecules of

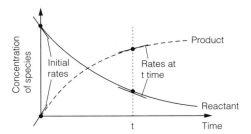

*Fig. 3.   The instantaneous rate of a reaction for a species is the slope of the tangent to the curve of concentration against time.*

reagent and products in the balanced chemical equation). For example, at any point in the reaction between hydrogen and nitrogen to form ammonia:

$$N_2 (g) + 3H_2(g) \rightarrow 2NH_3(g)$$

the rate of consumption of hydrogen is three times the rate of consumption of nitrogen, whilst the rate of production of ammonia is twice the rate of consumption of nitrogen but only two-thirds the rate of consumption of hydrogen.

**Rate law**

The **rate law** is the empirical relationship which describes the observed rate of reaction in terms of the concentrations of species in the overall reaction, including possibly the concentration of the products.

It is often observed that the rate of reaction is proportional to the product of the individual concentrations of the reactants raised to a simple power, for example, rate $\propto [A]^{\alpha}[B]^{\beta}$. Rate laws are empirical observations and do not necessarily fit the simple stoichiometry of the balanced chemical equation for the observed reaction but may be the consequence of a more complex underlying molecular reaction mechanism. For example, the apparently straightforward chemical reaction:

$$H_2(g) + Br_2(g) \rightarrow 2HBr(g)$$

has the experimentally determined rate law (see Topic F6):

$$\text{rate of formation of HBr} = \frac{k[H_2][Br_2]^{\frac{3}{2}}}{[Br_2] + k'[HBr]}$$

**Rate constants**

**Rate constants**, $k$, are the constants of proportionality which appear in the empirical **rate law** linking rate of reaction and concentration of species. The dimensions of the units of $k$ are dependent on the formulation of the individual rate law but can always be derived by dimensional analysis of the rate law. Thus a reaction which is **second order** overall must have a rate constant with dimensions of concentration$^{-1}$.time$^{-1}$ in order to provide the right hand side of the rate law with dimensions equal to the dimensions of concentration.time$^{-1}$ for rate of reaction. The exact units of $k$ depend on the units of concentration and time used, but mol dm$^{-3}$ and s, respectively, are common.

A rate constant for a particular reaction has a fixed value at a particular temperature, although it usually varies with temperature and the temperature dependence is often conveniently described by an **Arrhenius equation** (see Topic F3). Rate constants of **elementary reactions** do not vary with pressure so the observation of a pressure dependence in the rate of reaction indicates a more **complex** multistep reaction mechanism (see Topics F4, F5 and F6).

**Order of reaction**

If the rate law for a reaction can be written in the form, rate $\propto [A]^{\alpha}[B]^{\beta}...$ then the reaction is classified as $\alpha$-order in A, $\beta$-order in B,... and as $(\alpha+\beta+...)$-order overall. Where the exponent, or sum of exponents, equals one the reaction is said to be **first order** with respect to that species, or first order overall, respectively. Where the exponent, or sum of exponents, equals two the reaction is described as **second order** with respect to that species, or second order overall, respectively, and so on. Both the rate laws:

$$rate = k_1[A]^2$$

and

$$rate = k_2[A][B]$$

are second order overall, but whereas the first rate law is second order in species A only, the second rate law is first order in each of species A and B.

If a reactant species appears in the balanced chemical equation for the reaction but does not appear in the rate law then the reaction is **zero order** with respect to that species. Zero order terms are not usually written in rate law equations since the concentration of any species to the power zero is just unity. For example, the rate law for the aqueous phase iodination of propanone:

$$I_2 + H^+ + CH_3COCH_3 \rightarrow CH_2ICOCH_3 + HI + H^+$$

is:

$$rate = k[H^+][CH_3COCH_3]$$

The reaction is therefore zeroth order in iodine concentration, first order in each of the hydrogen ion and propanone concentrations, and second order overall.

The exponents do not have to be integers, and for rate laws not of the general form $[A]^{\alpha}[B]^{\beta}...$ the order is not a definable quantity. The rate law for the formation of HBr from $H_2$ and $Br_2$ is:

$$rate \text{ of formation of HBr} = \frac{k[H_2][Br_2]^{\frac{3}{2}}}{[Br_2] + k'[HBr]}$$

so the reaction is first order with respect to $H_2$ concentration, but has an indefinite order with respect to both $Br_2$ and HBr concentrations and an indefinite order overall (see Topic F6).

**Molecularity**

The **molecularity** of a reaction is the number of species which come together in the reaction. The complex rate law for the $H_2 + Br_2$ reaction indicates that the reaction does not proceed through a single step collision between undissociated hydrogen and bromine molecules, but consists of several separate **elementary reactions**. One of these is the reaction between hydrogen atoms and bromine molecules:

$$H + Br_2 \rightarrow HBr + Br$$

which come together to form a **collision complex** H-Br-Br that then breaks apart to form the product species HBr and Br. This is an example of a **bimolecular** reaction since two species are involved, H and $Br_2$.

A **unimolecular** reaction occurs when a single molecule acquires the necessary energy to break apart or rearrange its constituent atoms, for example, the thermal dissociation of azomethane:

$$CH_3N_2CH_3(g) \rightarrow 2CH_3(g) + N_2(g)$$

Except in the case of a true elementary reaction the molecularity is independent of the order. The order of a reaction is based entirely on experimental deduction of a rate law. In the example of the thermal decomposition of azomethane there are a number of hidden elementary reactions which determine the proportion of azomethane molecules which acquire sufficient energy to undergo unimolecular dissociation and the overall order of the reaction is not well defined (see Topic F5).

# F2 RATE LAW DETERMINATION

## Key Notes

**Method of isolation**

When the concentration of all other reactants is in large excess to the reactant under study, the concentration of the excess reactants can be assumed to remain constant as the reaction progresses and the order of the reaction with respect to the isolated reactant determined by direct observation of its concentration change with time. The method is commonly applied to convert second order reactions into pseudo-first order reactions.

**Method of initial rates**

A differential rate law of the general form $d[A]/dt = k[A]^\alpha[B]^\beta$ ... can be written as $\log |d[A]_0/dt| = \log k + \alpha \log[A]_0 + \beta \log[B]_0 +...$ for initial reagent concentrations $[A]_0$, $[B]_0$,... so the rate constant and order with respect to A can be determined from the intercept and gradient of a plot of the logarithm of the initial rate of reaction against $[A]_0$, for constant $[B]_0$.

**Integrated rate laws**

An integrated rate law expresses kinetic behavior directly in terms of the measurable quantities of concentration and time rather than instantaneous reaction rates.

**Integrated rate law: zeroth order reactions**

The integrated rate law of a reaction that is zeroth order with respect to removal of A is $kt = [A]_0 - [A]$. A plot of $[A]$ against $t$ is linear with gradient $-k$.

**Integrated rate law: first order reactions**

The integrated rate law of a reaction that is first order with respect to removal of A is $kt = \ln[A]_0 - \ln[A]$. A plot of $\ln[A]$ against $t$ is linear with gradient $-k$.

**Integrated rate law: second order reactions**

The integrated rate law of a reaction that is second order with respect to removal of A is $kt = 1/[A] - 1/[A]_0$. A plot of $1/[A]$ against $t$ is linear with gradient $k$.

**Half-lives**

The half-life, $t_{1/2}$, of a reaction is the time taken for the concentration of reactant to fall to half the initial value. The $t_{1/2}$ of reactions that are zero, first and second order with respect to removal of A are $[A]_0/2k$, $\ln2/k$ and $1/k[A]_0$, respectively, and the dependence of $t_{1/2}$ on the initial concentration can be used to determine the order of the reaction. $t_{1/2}$ of a first order reaction is independent of initial concentration.

**Related topics**

Empirical approaches to kinetics (F1)

Formulation of rate laws (F4)

Rate laws in action (F5)

The kinetics of real systems (F6)

## Method of isolation

The experimental determination of the **rate law** is considerably simplified by the **method of isolation** in which all the reactants except one are present in large excess. To a good approximation, the concentrations of the excess reactants remain constant during the reaction which enables the order of the reaction with respect to the **isolated reactant** to be determined directly from observation of the kinetics of just the isolated species. For example, if the true rate law for a reaction is:

$$\text{rate} = k\,[A][B]^2$$

and reactant B is in excess, then the concentration of B throughout the reaction can be approximated by its initial value $[B]_0$ and the rate law becomes:

$$\text{rate} = k'\,[A]$$

where $k' = k\,[B]_0^2$ is still a constant. Since the original **third order** reaction has been converted into a first order form the latter rate law is classified as **pseudo-first order** to indicate that the rate law disguises intrinsic higher order and only applies under particular conditions of reactant relative concentrations. $k'$ is called the pseudo-first order rate constant.

Similarly, if, instead, reactant A is present in large excess, the rate law becomes **pseudo-second order**:

$$\text{rate} = k''\,[B]^2$$

where $k'' = k\,[A]_0$ is the pseudo-second order rate constant. Pseudo rate laws of lower order, and involving only one species, are easier to identify and analyze than the complete law.

**Method of initial rates**

A **differential rate law** is the basic mathematical formulation of a rate law expressing the rate of change of a species concentration with time. It has the general form:

$$-\frac{d[A]}{dt} = k[A]^\alpha[B]^\beta\ldots$$

where $\alpha, \beta,\ldots$ is the **order** of reaction with respect to species A, B,… A positive sign to the differential of a particular species indicates rate of formation of that species, whereas a negative sign indicates rate of removal of that species.

If $[A]_0$, $[B]_0$… are the initial concentrations of species A, B,… then applying logarithms to this generalized differential rate law at time $t = 0$ gives:

$$\log\left|\frac{d[A]_0}{dt}\right| = \log k + \alpha \log[A]_0 + \beta \log[B]_0 + \ldots$$

The order, $\alpha$, of the reaction with respect to A is obtained from the slope of the graph of logarithm of **initial rate** of reaction of A against the logarithm of the corresponding starting concentration $[A]_0$ whilst values of $[B]_0,\ldots$ are held constant. Similarly, the value of $\beta$ is obtained by varying only $[B]_0$ and measuring the initial rate of reaction of B.

In the simpler situation in which the rate law involves only a single species:

$$-\frac{d[A]}{dt} = k[A]^\alpha$$

the logarithmic form (for all times) is:

$$\log\left|\frac{d[A]}{dt}\right| = \log k + \alpha \log[A]$$

and it is possible to obtain both the order and the rate constant for the rate law directly by plotting $\log|d[A]/dt|$ against $\log[A]$ values derived from a single experimental graph of variation in [A] with time (*Fig. 1*).

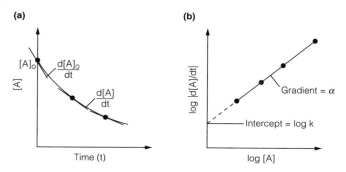

Fig. 1.    (a) Derivation of rates of reaction from tangents to the plot of concentration against time. (b) A plot of log rate vs. log concentration gives a straight line with gradient equal to the order, and intercept equal to log rate constant.

The determination of a rate law using the method of initial rates has a number of disadvantages:

(i)   determining tangents to a concentration *versus* time plot is generally subject to considerable uncertainty.

(ii)  the rate constant is obtained by extrapolation to an intercept which increases the error in $k$.

(iii) observing only initial rates may be misleading if the rate of reaction is also affected by the formation of products. The rate constant and order may be correct at times close to $t = 0$ but may not be valid over the whole course of the reaction. An example is the reaction between hydrogen and bromine to form HBr which has the rate law (see Topic F6) of:

$$\text{rate of formation of HBr} = \frac{k[H_2][Br_2]^{\frac{3}{2}}}{[Br_2] + k'[HBr]}$$

Initially, when [HBr] is small, rate $\propto [H_2][Br_2]^{1/2}$ but as the reaction progresses the significance of the $k'[HBr]$ term increases and the order with respect to $Br_2$ becomes undefined.

**Integrated rate laws**

Reaction rates are rarely measured directly because of the difficulty in determining accurate values for slopes of graphs. Instead an **integrated rate law** may be used which expresses kinetic behavior directly in terms of the measurable observables of concentration and time. Analytical expressions for integrated rate laws of simple types of reaction are presented in *Table 1* but even the most complex rate laws can usually be integrated numerically by computers. *Table 1* contains expressions of integrated rate laws in terms of concentration of reactant A at time $t$ but similar analytical expressions are readily derived for concentration of product P at time $t$. The advantage of these simple integrated rate laws is that the order of reaction with respect to a species is readily tested by means of a suitable plot of species concentration and time (*Fig. 2*).

*Table 1.   Integrated rate laws for reactions of simple order*

| Order | Reaction type | Differential rate law | Integrated rate law | Straight line plot | Half-life | Dimensions of $k$ |
|---|---|---|---|---|---|---|
| 0 | $A \rightarrow P$ | $\dfrac{d[A]}{dt} = -k$ | $kt = [A]_0 - [A]$ | $[A]$ vs. $t$ | $\dfrac{[A]_0}{2k}$ | conc. time$^{-1}$ |
| 1 | $A \rightarrow P$ | $\dfrac{d[A]}{dt} = -k[A]$ | $[A] = [A]_0 e^{-kt}$ <br> or, $kt = \ln\left(\dfrac{[A]_0}{[A]}\right)$ | $\ln[A]$ vs. $t$ | $\dfrac{\ln 2}{k}$ | time$^{-1}$ |
| 2 | $A \rightarrow P$ | $\dfrac{d[A]}{dt} = -k[A]^2$ | $kt = \dfrac{1}{[A]} - \dfrac{1}{[A]_0}$ | $\dfrac{1}{[A]}$ vs. $t$ | $\dfrac{1}{k[A]_0}$ | conc.$^{-1}$ time$^{-1}$ |
| 2 | $A + B \rightarrow P$ | $\dfrac{d[A]}{dt} = \dfrac{d[B]}{dt} = -k[A][B]$ | $kt = \dfrac{1}{([B]_0 - [A]_0)}\ln\left(\dfrac{[A]_0[B]}{[B]_0[A]}\right)$ | $\ln\left(\dfrac{[B]}{[A]}\right)$ vs. $t$ | See text | conc.$^{-1}$ time$^{-1}$ |

**Integrated rate law: zeroth order reactions**

A reaction which is **zeroth order**, or **pseudo-zeroth order** in removal of A, has a rate law of the form:

$$-\frac{d[A]}{dt} = k$$

where $k$ is the zeroth order rate constant (or pseudo-zeroth order rate constant if the **method of isolation** has been used). Separation of the variables and integrating:

$$\int d[A] = -k \int dt$$

with the condition that $[A] = [A]_0$ at $t = 0$ gives:

$$[A]_0 - [A] = kt$$

Thus a zeroth order reaction is identified by linearity in a plot of $[A]$ against $t$ and the gradient of the plot equals $-k$ (*Table 1*). The rate of removal of A is independent of $[A]$ as long as some A remains present. An example of such a reaction is the catalytic decomposition of ammonia, $NH_3$, at high concentrations, on hot tungsten. The observed zeroth order behavior is a consequence of decomposition of molecules of $NH_3$ adsorbed to the tungsten surface according to the **Langmuir adsorption isotherm** (see Topic F5).

**Integrated rate law: first order reactions**

The rate of removal of reactant A in a **first order** (or **pseudo-first order**) reaction is given by:

$$-\frac{d[A]}{dt} = k[A]$$

Separating the variables and integrating:

$$\int \frac{d[A]}{[A]} = -k \int dt$$

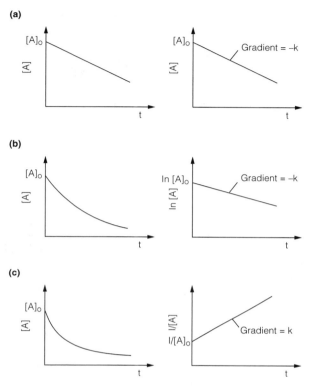

Fig. 2. The observed time dependence in concentration during removal of reactant A by (a) zeroth order, (b) first order, (c) second order kinetics. The second graph in each pair illustrates the linear plot of the corresponding rate law.

with the condition $[A] = [A]_0$ at $t = 0$ gives:

$$\ln[A] - \ln[A]_0 = -kt$$

The equation can also be written in the form:

$$[A] = [A]_0 e^{-kt}$$

which emphasizes the fact that all first-order reactions are characterized by an **exponential decay** of reactant concentration with time. The larger the value of the first order rate constant, $k$, the faster the decay in time. Since the exponent must be dimensionless all first-order rate constants have units of time$^{-1}$. A first order reaction is identified by linearity in the plot of $\ln[A]$ against $t$ (*Table 1*) and the gradient of this plot equals $-k$. Many chemical reactions and other physical processes are characterized by first order behavior (for example, radioactive decay), and many bimolecular reactions can be made to exhibit pseudo-first order behavior by ensuring one reactant is in excess. The advantage of first order kinetics is that the value of the rate constant can be derived from using only a relative measure of the concentration of A with time. The absolute concentration of A is not required.

**Integrated rate law: second order reactions**

The differential equation for a rate law that is **second order** (or **pseudo-second order**) in removal of species A is:

$$-\frac{d[A]}{dt} = k[A]^2.$$

Separating the variables and integrating:

$$\int -\frac{d[A]}{[A]^2} = k\int dt$$

with the condition that $[A] = [A]_0$ at $t = 0$ gives:

$$\frac{1}{[A]} - \frac{1}{[A]_0} = kt$$

The test for a second-order reaction is linearity in a plot of $1/[A]$ against $t$. The second-order rate constant is equal to the gradient of this plot (*Table 1*).

The analysis of a second-order reaction is slightly more complicated in the general case of a rate law that incorporates first-order removal in two separate species A and B (of equal stoichiometry) of different initial concentration $[A]_0$ and $[B]_0$, i.e.

$$-\frac{d[A]}{dt} = -\frac{d[B]}{dt} = k[A][B]$$

For reactions of this type a straight line is obtained from a plot of $\ln([B]/[A])$ against $t$ (*Table 1*) and the gradient of the line corresponds to $k([B]_0-[A]_0)$. However, to undertake such an analysis requires measurement of the absolute concentrations of both A and B simultaneously so, where possible, experimental conditions are arranged such that either $[A]_0 = [B]_0$, in which case the mathematics of the kinetics is the same as the second order integration in A given above, or that $[B]_0 \gg [A]_0$ (or *vice versa*), in which case the reaction reduces to pseudo-first order kinetics. The second approach yields the pseudo-first order rate constant, $k' = k[B]_0$, without needing to know absolute concentrations (from the plot of the natural logarithm of relative concentration of A with time) but the true second order rate constant is only obtained by repeating measurements

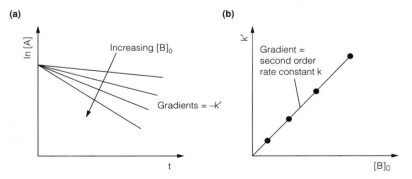

*Fig. 3.  Determination of a two-species second order rate constant, k, entails (a) pseudo-first order plots for species A followed by (b) a second order plot of pseudo first order rate constants k' against the corresponding excess initial concentration of species B.*

of $k'$ with different known initial concentrations of the excess reactant B (*Fig. 3*). This pseudo-first order approach is used extensively for the determination of second order rate constants of bimolecular elementary reactions.

**Half-lives**      The **half-life**, $t_{1/2}$, of a reaction is the time take for the concentration of reactant A to fall to half its value. Expressions for $t_{1/2}$ are obtained by substituting $[A] = [A]_0/2$ and $t = t_{1/2}$ into the **integrated rate law**. The corresponding half-life expressions are listed in *Table 1*. The half-life is always inversely proportional to $k$, but its dependence on $[A]_0$ depends on the order of the reaction. The half-life relationships may be used to determine the order of a reaction and its rate coefficient in two ways:

(i)   The reaction is followed over several half-lives and the dependence of the sequence of half-lives on the concentration at the start of each half-life period is examined.

(ii)  Several experiments can be performed, each with a different initial concentration, and the dependence on concentration of the first half-life for each experiment measured.

A useful diagnostic for first order reactions is that $t_{1/2}$ is independent of initial concentration, i.e. the concentration of A will fall to $[A]/2$ in an interval $\ln2/k$ whatever the value of $[A]$.

# F3 ENERGETICS AND MECHANISMS

## Key Notes

**Arrhenius equation**

The temperature dependence of the rate constant of the majority of chemical reactions is described by the Arrhenius equation, $k = Ae^{-E_a/RT}$ where $E_a$ (the activation energy) and A (the pre-exponential factor) are characteristic parameters for the reaction. They may be determined experimentally from a plot of $\ln k$ against $1/T$.

**Collision theory**

This simple model to describe the rate of a bimolecular reaction assumes that reaction occurs when two reactant species collide with an energy along their line of centers greater than the activation energy for the reaction. The species are treated as hard, structureless spheres that only interact when the distance between their centers is less than the collision radius (the sum of the radii of the colliding reactants). The derived rate constant also includes a steric factor to account for the probability that molecules collide with the correct relative orientation to permit reaction.

**Activated complex (transition state) theory**

This theory interprets chemical reaction in terms of a loosely-bound activated complex which acts as if it is in equilibrium with the reactant species. The molecular configuration of the activated complex corresponding to the maximum energy along the reaction coordinate between breaking of old bonds and formation of new bonds is known as the transition state. The derived rate constant is given by $k^{\ddagger}K^{\ddagger}$ where $K^{\ddagger}$ is the equilibrium constant between reactants and activated complex and $k^{\ddagger}$ is the first order rate constant for decomposition of the activated complex into products. These parameters can be calculated from statistical mechanics given a postulated model of the activated complex.

**Catalysts**

A catalyst increases the rate of chemical reaction by providing an alternative reaction pathway with lower activation energy than the reaction pathway in its absence. A catalyst is not consumed and therefore does not appear in the chemical equation for the reaction. A homogeneous catalyst is in the same phase as the reactants whilst a heterogeneous catalyst is in a different phase.

**Related topics**

Molecular behavior in perfect
  gases (A2)
Free energy (B6)
Fundamentals of equilibria (C1)

Empirical approaches to kinetics
  (F1)
Statistical thermodynamics (G8)

**Arrhenius equation**

The **rate constant**, $k$, and hence the **rate** of a chemical reaction, are usually observed to vary with temperature, $T$. For the majority of chemical reactions the rate constant increases with temperature. The temperature dependence is summarized mathematically in the **Arrhenius equation**:

$$k = Ae^{\frac{-E_a}{RT}}$$

The two parameters $A$ and $E_a$ are together known as the **Arrhenius parameters** and are characteristic to each reaction. The parameter $A$ has the same units as $k$ and is called the **pre-exponential** or **Arrhenius factor**, and $E_a$ is called the **activation energy**. ($R$ is the **universal gas constant**.) The Arrhenius equation can also be written in the form:

$$\ln k = \ln A - \frac{E_a}{RT}$$

Therefore, a plot of $\ln k$ against $1/T$ produces a straight line with slope equal to $-E_a/R$ and intercept equal to $\ln A$.

Reaction rate increases with temperature when $E_a$ is positive (which is generally the case). The larger the activation energy the greater is the sensitivity of the reaction to changes in temperature. A reaction with an activation energy close to zero has a rate that is largely independent of temperature. Most reactions have an activation energy somewhere in the range of a few tens to a few hundreds of kJ mol⁻¹ and a useful rule of thumb is that reactions with $E_a$ in the range 50–60 kJ mol⁻¹ have rate constants that approximately double for each 10 K rise in temperature at around room temperature.

A reaction with a negative activation energy (corresponding to the observation of a decrease in rate with increase in temperature) usually indicates that the observed rate constant is a composite of rate constants of **elementary reactions** contributing to a **complex mechanism**. For example, if $k = k_1k_2/k_3$ and $k_3$ increases more rapidly with temperature than the product $k_1k_2$ then $k$ will decrease overall (see Topic F4).

**Collision theory**

**Collision theory** is a theoretical framework to explain the origin of the **Arrhenius equation**. The fundamental assumption of collision theory is that reaction occurs when two molecules collide with one another in a **bimolecular** reaction. As in the **kinetic theory of gases** (see Topic A2) collision theory makes the assumption that molecules are hard, structureless spheres (like billiard balls) which do not interact until they come into direct contact. It is further assumed that reaction only occurs when molecules collide with a kinetic energy greater than some threshold value.

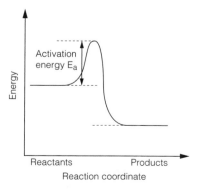

*Fig. 1. A reaction profile. The activation energy is the height of the barrier above the energy of the separated reactants.*

The basis for this assumption is the **reaction profile** (*Fig. 1*) which plots the change in **energy** of the molecules as the reaction proceeds from reactants to products. The energy rises as the separation of the molecules becomes small enough for them to be in contact because bonds distort and break. To the right of the maximum in the profile the energy decreases as new bonds form and the product molecules separate to distances where there is no longer any interaction. The height of the energy barrier from reactants to products is identifiable with the **activation energy**, $E_a$, of the reaction. Molecules that collide with kinetic energy less than $E_a$ bounce apart without reaction.

According to the Boltzmann distribution (see Topic G8) the number of molecules which have an energy greater than an energy $E_a$ is proportional to the factor, $e^{-E_a/RT}$. Since the rate of collision between molecules A and B is directly proportional to the concentration of A and B, it follows that the rate of collisions with energy > $E_a$ is given by:

$$\text{rate of reaction} \propto [A][B]e^{-E_a/RT}$$

Comparison of this expression with the second order **rate law** for reaction between A and B:

$$\text{rate of reaction} = k[A][B]$$

shows that the observed second order rate constant:

$$k \propto e^{-\frac{E_a}{RT}}$$

which is exactly the form of the **Arrhenius equation** with the constant of proportionality identifiable as the **pre-exponential Arrhenius factor** $A$.

The value of $A$ can be calculated using the kinetic theory of gases by assuming that two hard spheres collide when the distance between them is less than the sum of their radii. A molecule of A travelling at speed $s$ through a concentration [B] of B molecules will collide with all B molecules that lie within the cylinder of radius $d$ around A's trajectory, where $d = r_A + r_B$ and $r_A$, $r_B$ are the radii of molecules A and B (*Fig. 2*). Since the volume of the cylinder swept out per unit time is equal to $\pi d^2 s$ the number of collisions per unit time per unit volume for a single molecule of A with molecules of B is $\pi d^2 s N_A[B]$, where $N_A$ is Avogadro's number. So for a reaction mixture containing a concentration [A] of A molecules the total rate of collisions per unit time per unit volume is $\pi d^2 s N_A[B][A]$. (The quantities $d$ and $\pi d^2$ are called, respectively, the **collision radius** and **collision cross-section**.)

In reality, molecules in a gas at temperature $T$ do not have a single speed but

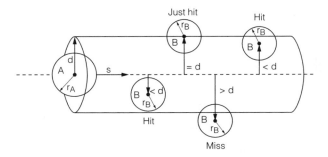

*Fig. 2.   The collision volume swept out by a molecule of A passing through stationary molecules of B.*

a range of speeds described by the Maxwell-Boltzmann distribution (see Topic A2). Therefore the **mean relative speed**:

$$\bar{s} = \left( \frac{8k_B T}{\pi \mu} \right)^{\frac{1}{2}}$$

must be used for $s$ in the expression for collision rate just derived. (The reduced mass $\mu = m_A m_B / (m_A + m_B)$ occurs in the equation for $\bar{s}$ since what matters is the relative speed of approach of the molecules.) When the Boltzmann factor is included, the collision theory expression becomes:

$$\text{rate of reaction} = \pi d^2 \bar{s} N_A [A][B] e^{-\frac{E_a}{RT}}$$

from which is obtained the pre-exponential Arrhenius factor, $A = \pi d^2 \bar{s} N_A$. The experimental value of $A$ is often smaller than the calculated value because molecules must also have specific orientation to each other at the moment of collision as well as sufficient kinetic energy. This is accounted for by including a **steric factor, $P$,** in the pre-exponential factor, $A = P \pi d^2 \bar{s} N_A$, to represent the probability that an encounter has the correct orientation to permit chemical reaction (*Fig. 3*). The value of $P$ lies between 0 (no relative orientations lead to reaction) and 1 (all relative orientations lead to reaction).

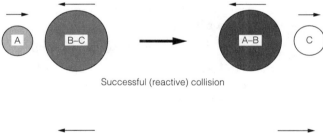

Successful (reactive) collision

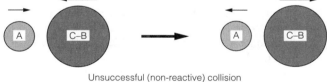

Unsuccessful (non-reactive) collision

Fig. 3.  *The role of specific orientation of reactants in determining a reactive (or non-reactive) outcome of a collisional encounter.*

**Activated complex (transition state) theory**

The **activated complex theory** of chemical reaction assumes that the maximum in the energy curve of the reaction profile (*Fig. 2*) corresponds to the formation of an **activated complex** which has a definite, loosely bound, structure with maximum distortion of bonds (*Fig. 4*). At this point the atoms or molecules are in the **transition state** between breakage of old bonds and formation of new bonds. The path along which reactants come together to pass through the transition state and separate into products is called the **reaction co-ordinate**.

An activated complex, $AB^{\ddagger}$ behaves as if it is in equilibrium with its reactants:

$$A + B \rightleftharpoons AB^{\ddagger} \rightarrow \text{products}$$

with concentration described in the normal way (neglecting **activity coefficients**) by an **equilibrium constant, $K^{\ddagger}$,** (Topic C1):

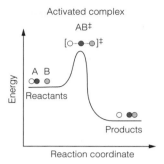

*Fig. 4.   The activated complex interpretation of a reaction profile.*

$$K^{\ddagger} = \frac{\left[AB^{\ddagger}\right]}{[A][B]}$$

The **rate** at which products form is proportional to the concentration of the activated complex so:

$$\text{rate of reaction} = k^{\ddagger}[AB^{\ddagger}] = k^{\ddagger}K^{\ddagger}[A][B]$$

where $k^{\ddagger}$ is the **first order rate constant** associated with decomposition of the activated complex. Comparison of this expression with the second order **rate law** for the reaction between A and B:

$$\text{rate of reaction} = k[A][B]$$

shows that the observed reaction rate constant, $k = k^{\ddagger}K^{\ddagger}$.

Values for the $k^{\ddagger}$ and $K^{\ddagger}$ constants are often calculated by **statistical mechanics** (see Topic G8) using values for bond lengths and bond frequencies in the postulated structure of the activated complex. The resulting expressions describe the rate constant in terms of how the **partition functions** of translation, vibration and rotation modes change from those of the isolated reactants to those of the activated complex.

Additional insight into the physical basis of chemical reaction is obtained by applying a thermodynamic formulation to activated complex theory. Equilibrium thermodynamics shows that an equilibrium constant can be written in terms of the standard **Gibbs free energy** (see Topics B6 and C1), which in this case is an **activation Gibbs free energy**, $\Delta G^{\ddagger}$, for formation of the activated complex, i.e.

$$K^{\ddagger} = e^{-\frac{\Delta G^{\ddagger}}{RT}}$$

Since, $\Delta G^{\ddagger} = \Delta H^{\ddagger} - T\Delta S^{\ddagger}$, the observed reaction rate constant can be written as:

$$k \propto e^{-\frac{(\Delta H^{\ddagger} - T\Delta S^{\ddagger})}{RT}} \propto e^{\frac{\Delta S^{\ddagger}}{R}} e^{-\frac{\Delta H^{\ddagger}}{RT}}$$

This expression also has the form of the **Arrhenius equation** when the **enthalpy of activation**, $\Delta H^{\ddagger}$, is identified with the **activation energy** $E_a$, and the **entropy of activation**, $\Delta S^{\ddagger}$, is identified with the **pre-exponential factor** (or more precisely with $R\ln A$). For a reaction which has strict orientation requirements (for example the approach of a substrate molecule to an enzyme) the entropy of activation will be strongly negative (because of the decrease in disorder when the activated complex forms) and the pre-exponential factor will be small compared with a reaction that does not have such strict orientation requirements. Thus activated complex theory incorporates information about the intrinsic geometry of the transition state to account for the **steric factor**, $P$, arbitrarily introduced into the pre-exponential factor derived from **collision theory**.

**Catalysts**

The **rate constant** for a reaction depends on the temperature, the height of the **activation barrier** $E_A$ and the magnitude of the **pre-exponential Arrhenius factor** $A$. Whilst increasing the temperature can be used to increase the rate constant the latter two parameters cannot be altered since they are specific to the particular reaction path and determined by the electronic structure and bonding arrangement of the reactants and **activated complex**. Instead, a **catalyst** may be available that increases the rate of reaction by providing an alternative reaction path with a lower activation energy (*Fig. 5*) so that at a given temperature a greater proportion of collisions have energy greater than the activation energy. (Note that the rate of back reaction must also increase when the height of the activation barrier is lowered.)

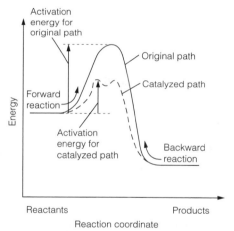

*Fig. 5.* *The energetics of the reaction path between reactants and products for a reaction with and without a catalyst.*

A catalyst merely provides an alternative reaction mechanism and is not consumed in the reaction. Furthermore, a catalyst does not affect the equilibrium distribution between reactants and products, which is determined solely by **thermodynamics**, but increases the rate at which the equilibrium is reached.

Catalysts which occupy the same phase as the reactants (for example all in solution) are called **homogeneous catalysts** whilst those which are of a different phase (for example a solid surface in a gas or liquid phase reaction) are called **heterogeneous catalysts**. The latter are particularly prevalent in industrial processes. The alternative reaction mechanism is usually provided by **physisorption** or **chemisorption** of one or more reactants onto the surface which may weaken certain bonds and enhance the chance of close encounter. An important class of natural homogeneous catalysts are **enzymes** (Topic F6).

# **F4** FORMULATION OF RATE LAWS

---

## Key Notes

| | |
|---|---|
| **Rate laws of elementary reactions** | An elementary reaction is a single reaction step. When only a single molecule is involved (A → P) the elementary reaction is unimolecular with a first order rate law (rate = $k$[A]). If two reactant molecules are involved (A + B → P) the elementary reaction is bimolecular with a second order rate law (rate = $k$[A][B]). |
| **Complex reaction** | A complex reaction is one which proceeds through more than one constituent elementary reaction step. Unimolecular reactions, chain reactions, catalytic and enzyme reactions are all examples of complex reactions. |
| **Steady state assumption** | This is the assumption that the concentrations of all intermediate species in a reaction mechanism remain constant during the reaction. Hence the net change in concentration [I] of any intermediate with time can be set to zero, $d[I]/dt \approx 0$ which means that the rates of formation and removal for each intermediate must balance. |
| **Formulating rate laws** | The overall rate law of a complex reaction mechanism is formulated by combining the first and second order rate laws of the constituent elementary reactions, usually by applying the steady state assumption or, analogously, by assuming that some equilibrium is attained. The formulated rate law must be consistent with the observed rate law. |
| **Rate determining step** | The rate determining step is the slowest step in a reaction mechanism. The rate of this reaction determines the maximum overall rate of formation of products. |

| **Related topics** | Empirical approaches to kinetics (F1) | Rate laws in action (F5) |
|---|---|---|
| | Rate law determination (F2) | The kinetics of real systems (F6) |

---

**Rate laws of elementary reactions**

An **elementary reaction** is a single, discrete reaction step. The **molecularity** of the reaction is the number of reactant molecules involved in this discrete reaction step (see Topic F1). The **rate law** of an elementary **unimolecular** reaction (i.e. A → P) is intrinsically first order in the reactant since the rate of reaction at any time is proportional only to the concentration of A remaining:

$$\text{rate} = k[A]$$

The rate law of an elementary **bimolecular** reaction (i.e. A + A → P, or A + B → P) is intrinsically second order since the rate of reaction at any time is proportional to the rate of collision between the two molecules which in turn is directly proportional to the concentration of both molecules remaining:

$$\text{rate} = k[A]^2$$

or

$$rate = k[A][B]$$

**Complex reaction**    The term **complex reaction** is used for a reaction which consists of more than one constituent **elementary reaction** step. A complex reaction proceeds through the formation and removal of **intermediate species** not contained in the balanced chemical equation written for the reaction.

The overall rate law of a complex reaction is derived by combining the rate laws of the constituent elementary reactions and must be expressed only in terms of concentrations of reactants or products appearing in the overall balanced chemical equation for the reaction. It must also agree with the observed rate law under all sets of reaction conditions. The reverse situation, that of deducing complex behavior from an observed rate law is often not straightforward. For example, if an observed reaction is presumed to be genuinely bimolecular then its rate law is, by definition, second order. However, if the observed rate law is second order, then the reaction may be bimolecular, or it may be complex, and the latter might only be deduced after detailed investigation across a wide range of reaction conditions. This is the case for the reaction between $H_2$ and $I_2$:

$$H_2(g) + I_2(g) \rightarrow 2HI(g)$$

which shows second order kinetics:

$$rate = k[H_2][I_2]$$

and was presumed to be a bimolecular reaction between the two reactant diatomics. In fact there is an underlying **chain reaction mechanism** involving radical species, as for the reaction between $Br_2$ and $H_2$ (see Topic F6).

Complex reactions are abundant in chemistry. Examples include unimolecular dissociation (or rearrangement) reactions, enzyme or surface catalysis, and chain and explosion reactions (see Topics F5 and F6).

**Steady state assumption**    The **steady state assumption** presumes that the concentrations of all **intermediate species**, I, in the reaction mechanism remain constant and small during the reaction (except right at the beginning and right at the end), or in other words that the net change in concentration [I] with time is zero, $d[I]/dt \approx 0$. The assumption is equivalent to equating the total rate of removal of each intermediate with its total rate of formation.

Application of the steady state assumption effectively converts the differential equation for each intermediate into an algebraic expression from which the steady state concentrations can be derived for substitution into the overall rate law. The steady state assumption is widely applied in kinetics to formulate rate laws of **complex reactions**.

**Formulating rate laws**    The first step in formulating a **rate law** is to write down the proposed mechanism in terms of individual **unimolecular** and **bimolecular elementary reaction** steps. Since a rate law is an experimentally derived property of a reaction (see Topic F1) the combination of the individual first and second order rate laws must yield an overall rate law that is consistent with observation. For example, the gas phase oxidation of nitrogen monoxide, NO:

$$2NO(g) + O_2(g) \rightarrow 2NO_2(g)$$

is observed experimentally to be third order overall:

$$\text{rate of formation of } NO_2 = k[NO]^2[O_2]$$

Although one explanation for the third order behavior might be a single ter-molecular (three molecule) elementary reaction, the simultaneous collision between three molecules is extremely unlikely so the reaction is likely to be **complex**. The additional observation that rate of NO oxidation decreases with increasing temperature is further evidence of a complex reaction because rates of elementary reactions increase with temperature (Topic F3).

A reaction mechanism can be postulated which proceeds through the forma-tion of an $N_2O_2$ dimer, which may dissociate back into two NO molecules or undergo reactive collision with an $O_2$ molecule to produce two $NO_2$ molecules. The decomposition of $N_2O_2$ into NO is a unimolecular reaction step, the other two are bimolecular.

$$NO + NO \rightarrow N_2O_2 \qquad\qquad \text{rate of formation of } N_2O_2 = k_1[NO]^2$$

$$N_2O_2 \rightarrow NO + NO \qquad\qquad \text{rate of decomposition of } N_2O_2 = k_{-1}[N_2O_2]$$

$$N_2O_2 + O_2 \rightarrow NO_2 + NO_2 \qquad \text{rate of consumption of } N_2O_2 = k_2[N_2O_2][O_2]$$

(the subscripts on the rate constants merely represent labels for the corre-sponding reaction.) The overall rate at which $NO_2$ is formed is controlled by the third reaction in the mechanism:

$$\text{rate of formation of } NO_2 = 2k_2[N_2O_2][O_2]$$

(The factor 2 appears in this rate law because two $NO_2$ molecules are formed in each reactive collision encounter between $N_2O_2$ and $O_2$ so the rate of formation of $NO_2$ is twice the rate at which the collisions occur.)

The **steady state assumption** is used to remove from the rate law expression the term involving the concentration of the intermediate species $N_2O_2$. The net rate of formation of $N_2O_2$ is given by the difference between its rate of formation (*via* reaction 1) and its rate of removal (*via* reactions –1 and 2), and under the steady state assumption is approximated to zero:

$$\frac{d[N_2O_2]}{dt} = k_1[NO]^2 - k_{-1}[N_2O_2] - k_2[N_2O_2][O_2] \approx 0$$

Rearranging gives:

$$[N_2O_2] = \frac{k_1[NO]^2}{k_{-1} + k_2[O_2]}$$

so that the rate of formation of $NO_2$ in terms of reactants only is:

$$\text{rate of formation of } NO_2 = \frac{2k_1k_2[NO]^2[O_2]}{k_{-1} + k_2[O_2]}$$

This rate law is more complex than the observed rate law. However, if it is the case that the rate of decomposition of $N_2O_2$ (reaction –1) is much greater than its rate of reaction with $O_2$ (reaction 2), then $k_{-1}[N_2O_2] \gg k_2[N_2O_2][O_2]$, (i.e. $k_{-1} \gg k_2[O_2]$) and the formulated rate law becomes:

$$\text{rate of formation of } NO_2 \approx \left(\frac{2k_1k_2}{k_{-1}}\right)[NO]^2[O_2]$$

in agreement with the observed rate law. Furthermore, since the observed third order rate constant is actually a combination of elementary reaction rate constants:

$$k \approx \frac{2k_1 k_2}{k_{-1}}$$

the observed negative temperature dependence is explained if the rate of increase with temperature of rate constant $k_{-1}$ is more rapid than the rate of increase of the product $k_1 k_2$. (Note that in the general case of rate law formulation using similar procedures it may not be possible to invoke an approximation such as $k_{-1} \gg k_2[O_2]$ used above, in which case both terms must explicitly remain in the derived rate law.)

**Rate determining step**

The **rate determining step** is the slowest reaction in a reaction mechanism and consequently controls the rate of overall reaction to form products. In the rate law derived in the preceding section for the oxidation of NO:

$$\text{rate of formation of } NO_2 \approx \left(\frac{2k_1 k_2}{k_{-1}}\right)[NO]^2[O_2]$$

the rate of reaction of $N_2O_2$ with $O_2$ is assumed to be much slower than the rate of $N_2O_2$ decomposition (i.e. $k_{-1} \gg k_2[O_2]$). In other words, reaction between $N_2O_2$ and $O_2$ is the rate determining step in this case.

Under these conditions the overall rate of reaction is controlled by the rate constant $k_2$ and the concentration of $O_2$. However, if the concentration of $O_2$ in the reaction mixture is sufficiently large that $k_2[O_2] \gg k_{-1}$ the expression for the derived rate law becomes:

$$\text{rate of formation of } NO_2 \approx \frac{2k_1 k_2 [NO]^2 [O_2]}{k_2[O_2]} \approx 2k_1 [NO]^2$$

and the reaction is now second order in NO only. The concentration of $O_2$ is no longer important once $[O_2]$ exceeds a certain value because essentially all molecules of $N_2O_2$ that form react with $O_2$ to produce $NO_2$ before they have a chance to decompose back into NO. Consequently the rate of reaction is independent of $[O_2]$ and determined solely by the rate at which $N_2O_2$ is formed through reaction 1. These two scenarios illustrate that rate determining step is not necessarily a fixed entity but may switch from one elementary step to another for different experimental conditions.

# F5 RATE LAWS IN ACTION

## Key Notes

**Opposing reactions**
A system approaching equilibrium consists of a forward reaction from reactants to products and an opposing back reaction from products to reactants. The rate at which the system approaches equilibrium is equal to the sum of the forward and backward rates.

**Lindemann mechanism**
The Lindemann mechanism is the sequence of underlying elementary reaction steps that combine to yield an overall first order rate law for observed unimolecular reactions. The mechanism postulates that bimolecular collisions between molecules A produce activated intermediates A* which either deactivate through further collision or proceed along the reaction path to products.

**Langmuir surface adsorption kinetics**
For a gas of partial pressure $p$ above a surface with fractional coverage $\theta$ (the ratio of the number of surface sites occupied to the number available), rate of adsorption of gas to the surface is $k_1 p(1-\theta)$ and rate of desorption of gas from the surface is $k_{-1}\theta$, where $k_1$ and $k_{-1}$ are the adsorption and desorption rate constants, respectively. If the adsorbed gas undergoes unimolecular reaction with rate constant $k_2$ the observed overall rate of reaction is $k_1 k_2 p/(k_{-1} + k_1 p)$.

**Photochemical rate laws**
A photochemical reaction is initiated by absorption of one or more photons. The corresponding rate law is the product of the concentration of the absorbing species and a photochemical rate constant $J$ and is first order. The value of the rate constant incorporates terms for the intensity of incident light and the absorption coefficient of the molecule integrated over all appropriate wavelengths.

**Related topics**
Empirical approaches to kinetics (F1)
Rate law determination (F2)
Formulation of rate laws (F4)
The kinetics of real systems (F6)

**Opposing reactions**

All chemical reactions are potentially reversible but usually the reverse reaction is so slow that it can be neglected. However, for reactions approaching an equilibrium that is not overwhelmingly in favor of products, the back reaction converting products into reactants becomes important since the overall net rate of reaction must decrease (and is zero at equilibrium) (see Topic C1). Consider a general case of an isomerization reaction between A and B in which the opposing kinetics are **first order** (or **pseudo-first order**):

$$A \underset{k_{-1}}{\overset{k_1}{\rightleftharpoons}} B$$

and the equilibrium concentrations are $[A]_e$ and $[B]_e$. When the reaction is not at equilibrium the concentrations of A and B can be written as $[A]_e + x$ and $[B]_e - x$, where $x$ represents some arbitrary displacement in concentration (positive or negative) away from equilibrium. The net reaction towards equilibrium at this instant is described by the **rate law**:

$$\frac{d\left(\left[A\right]_{e}+x\right)}{dt}=-k_{1}\left(\left[A\right]_{e}+x\right)+k_{-1}\left(\left[B\right]_{e}-x\right)$$

or, on rearrangement, by:

$$\frac{d\left[A\right]_{e}}{dt}+\frac{dx}{dt}=-k_{1}\left[A\right]_{e}+k_{-1}\left[B\right]_{e}-\left(k_{1}+k_{-1}\right)x$$

At equilibrium, $x = 0$, and there is no net reaction so:

$$\frac{d\left[A\right]_{e}}{dt}=-k_{1}\left[A\right]_{e}+k_{-1}\left[B\right]_{e}=0$$

which on substitution above gives:

$$\frac{dx}{dt}=-\left(k_{1}+k_{-1}\right)x$$

This equation shows that the approach to equilibrium of opposing first order reactions is also a first order process with a first order rate constant equal to the sum of the forward and reverse first order rate constants, i.e. opposing reactions approach equilibrium at a rate faster than either the forward or backward reactions alone. The relaxation to equilibrium of a mixture initially containing concentration $[A]_0$ of A and zero concentration of B is shown in *Fig. 1*.

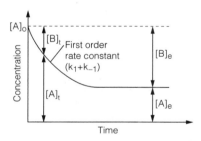

*Fig. 1.* Approach to equilibrium for the opposing reactions A $\underset{k_{-1}}{\overset{k_{1}}{\rightleftharpoons}}$ B starting from initial A and B concentrations of $[A]_0$ and 0, respectively.

**Lindemann mechanism**

The rate of observed apparent first order reactions, A → P, generally increases with temperature, indicating that the reactant must surmount an energy barrier, yet a **first order rate law** apparently excludes the possibility of achieving activation through collision. The **Lindemann mechanism** postulates a series of underlying elementary reaction steps contributing a **complex** reaction:

$$A+A\underset{k_{-1}}{\overset{k_{1}}{\rightleftharpoons}}A^{*}+A$$

$$A^{*}\xrightarrow{k_{2}}P$$

A reactant molecule, A, is excited to an energized state A* by collision with another reactant molecule A (reaction 1). A* may either be collisionally deactivated back to A (reaction –1) or continue along the reaction path to form product, P (reaction 2). The overall rate of formation of products is:

$$\frac{d[P]}{dt} = k_2[A^*]$$

The concentration of A* required for substitution into the rate law is obtained using the **steady state approximation** for A*, i.e. by equating the net rate of formation of A* to zero:

$$\frac{d[A^*]}{dt} = k_1[A]^2 - k_{-1}[A^*][A] - k_2[A^*] = 0$$

Therefore:

$$[A^*] = \frac{k_1[A]^2}{k_{-1}[A] + k_2}$$

which gives an overall observed rate of reaction of:

$$\frac{d[P]}{dt} = \frac{k_2 k_1[A]^2}{k_{-1}[A] + k_2}$$

Although this rate law is not first order, if the concentration of A is sufficiently high that the rate of deactivation collisions between A* and A is greater than the rate of unimolecular reaction of A*, then $k_{-1}[A^*][A] \gg k_2[A^*]$ (i.e. $k_{-1}[A] \gg k_2$) and the rate law simplifies to:

$$\frac{d[P]}{dt} = \frac{k_2 k_1}{k_{-1}}[A]$$

This expression is now a first order rate law in which the observed first order rate constant, $k_{uni}$, is a composite of rate constants for underlying elementary reactions:

$$k_{uni} = \frac{k_1 k_2}{k_{-1}}$$

The Lindemann mechanism is easily adapted to the situation in which activation of A is dominated by collisions with molecules of a non-reactive diluent bath gas, M, rather than other molecules of A. The same kinetic approach for the elementary reactions:

$$A + M \underset{k_{-1}}{\overset{k_1}{\rightleftharpoons}} A^* + M$$

$$A^* \overset{k_2}{\longrightarrow} P$$

gives a rate of product formation:

$$\frac{d[P]}{dt} = \frac{k_2 k_1[A][M]}{k_{-1}[M] + k_2}$$

At any given pressure the concentration of M is constant so the above rate law is equivalent to a first order rate law:

$$\frac{d[P]}{dt} = k_{uni}[A]$$

with

$$k_{uni} = \frac{k_1 k_2[M]}{k_{-1}[M] + k_2}$$

The Lindemann mechanism also explains another experimental feature of first order reactions, that the value of $k_{uni}$ varies with pressure. Towards the **high pressure limit** (where [M] is large and $k_{-1}[M] \gg k_2$) the value of $k_{uni}$ becomes independent of pressure:

$$k_{uni} = \frac{k_1 k_2}{k_{-1}}$$

The physical basis for this is that at sufficiently high pressures the rate of collisional activation of A to A* is sufficiently fast to always maintain equilibrium between the two so that the **rate determining step** for overall reaction is the first order elementary reaction from A* to P. Conversely, towards the **low pressure limit** (where $k_2 \gg k_{-1}[M]$), the reaction effectively becomes bimolecular:

$$\frac{d[P]}{dt} = k_1[M][A]$$

with the rate determining step being the rate at which the bimolecular collisions between A and M yield activated A*.

**Langmuir surface adsorption kinetics**

Suppose a surface consists of a number of energetically equivalent adsorption sites to which a gaseous reactant molecule A can bind reversibly:

A + site ⇌ A-site

and that $\theta$ denotes the fraction of all sites currently occupied. The rate of adsorption at that instant is proportional to the pressure of A in the gas, $p_A$, and to the fraction of sites currently unoccupied $(1 - \theta)$:

rate of adsorption = $k_1 p_A (1-\theta)$

where $k_1$ is the associated adsorption rate constant. The rate of desorption at the same instant is proportional to the fraction of sites occupied:

rate of desorption = $k_{-1} \theta$

where $k_{-1}$ is the desorption rate constant. At equilibrium the rates of adsorption and desorption must equal, so:

$$k_1 p_A (1 - \theta) = k_{-1}\theta$$

and

$$\theta = \frac{k_1 p_A}{k_{-1} + k_1 p_A}$$

This equation is known as the **Langmuir adsorption isotherm** (*Fig. 2a*). As the pressure increases from zero, $\theta$ rises, first linearly with $p_A$ since $k_{-1} \gg k_1 p_A$, but tends to unity (complete **monolayer** coverage) at high pressure when $k_1 p_A \gg k_{-1}$.

The Langmuir adsorption isotherm is readily adapted to describe the kinetics of unimolecular decomposition of a surface-adsorbed species. For example, ammonia ($NH_3$) decomposes on hot tungsten according to:

$NH_3(g) \rightleftharpoons NH_3(ads) \rightarrow$ products

The observed rate of decomposition of $NH_3$ is equal to $k_2[NH_3(ads)]$. So provided decomposition is sufficiently slow that the adsorption equilibrium is not disturbed, $[NH_3(ads)]$ is equal to the surface coverage $\theta$ and

$$\text{rate of reaction} = \frac{k_1 k_2 p_{NH_3}}{k_{-1} + k_1 p_{NH_3}}$$

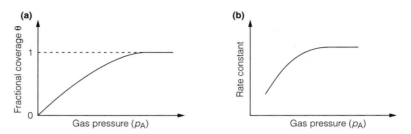

Fig. 2.  (a) The Langmuir adsorption isotherm for fraction surface coverage θ as a function of gas pressure, for monolayer coverage. (b) The variation with gas pressure of the rate constant of a unimolecular surface decomposition reaction.

Although the full rate law is complex, the reaction has two limiting rate laws corresponding to the two extremes of the Langmuir adsorption isotherm (*Fig. 2b*). At high pressure, when the surface is **saturated** (completely covered) and $\theta = 1$ (i.e. $k_1 p_{NH_3} \gg k_{-1}$),

$$\text{rate of reaction} = k_2$$

and the reaction is **zero order**; there is no dependence of rate on concentration of $NH_3$. At low pressure, when surface coverage $\theta$ is small (i.e. $k_{-1} \gg k_1 p_{NH_3}$),

$$\text{rate of reaction} = \frac{k_1 k_2}{k_{-1}} p_{NH_3}$$

and the reaction is **first order** in $NH_3$.

**Photochemical rate laws**

A photochemical reaction is one initiated by absorption of one or more photons of **electromagnetic radiation**. Obvious examples are reactions initiated by solar radiation such as the absorption of red and blue light by molecules of chlorophyll leading to production of carbohydrates through photosynthesis, or the absorption of ultraviolet light by molecules of oxygen in the upper atmosphere to produce Earth's protecting ozone layer (see Topic I7).

The **rate** of a photochemical elementary reaction is directly proportional to the concentration of absorbing species and is therefore described by **first order** kinetics. The constant of proportionality is called the **photochemical rate constant**, and usually given the symbol $J$. Using the photodissociation of $O_2$ into two O atoms as an example:

$$O_2 + h\nu \rightarrow 2O$$

the rate of removal of $O_2$ is:

$$-\frac{d[O_2]}{dt} = J[O_2]$$

and the **half-life** for $O_2$ removal is, $t_{1/2} = \ln 2 / J$.

The magnitude of the photochemical rate constant is a function of the intensity of the incident light causing the photochemical process and the intrinsic ability of the molecule to absorb photons (known as the **absorption coefficient**) for all appropriate wavelengths of incident light. Therefore the value of a solar photochemical rate constant varies with time of day, latitude and season, etc., because the intensity of solar radiation varies with these parameters. But for any particular set of irradiation conditions the constant $J$ may be treated analogously to first order thermal rate constants, $k$.

The **quantum yield**, $\Phi$, of a photochemical reaction is equal to the ratio of the number of molecules or radicals of the product under consideration to the number of photons absorbed:

$$\Phi = \frac{\text{number of particular product species produced}}{\text{number of photons absorbed}}$$

The absorption by an $O_2$ molecule of one photon of ultraviolet of appropriate wavelength produces two O atoms. Therefore the quantum yield for O atom formation *via* this photodissociation process is two.

# F6 THE KINETICS OF REAL SYSTEMS

## Key Notes

**Chain reactions**

A chain reaction occurs when a reaction intermediate generated in one step reacts with another species to generate another reaction intermediate. A chain reaction mechanism typically contains several types of elementary reaction steps including initiation, propagation, branching and termination. Mechanisms containing many branching reactions may lead to explosions.

**Explosions**

Chain branching explosions can arise when an elementary reaction produces more reaction intermediates than are consumed, each of which then instigate further chain branching reactions resulting in a catastrophic accelerated increase in reaction rate. Whether explosion or smooth reaction occurs depends on exact conditions of temperature and pressure.

**Enzyme kinetics: the Michaelis-Menten equation**

In the lock and key hypothesis of enzyme action, enzyme and substrate are in equilibrium with an enzyme-substrate complex (ES) which can proceed through to products. The Michaelis-Menten equation for the rate of formation of products is $v = k_2[E]_0[S]/(K_M + [S])$, where $[E]_0$ is total enzyme present and $K_M$ is the Michaelis constant. The enzyme released from the ES complex is available for further reaction and is therefore a catalyst.

**Enzyme kinetics: Lineweaver-Burke plots**

A Lineweaver-Burke plot is a linear relationship used to analyze kinetic data on enzyme catalyzed reactions. The reciprocal of the rate of reaction $(1/v)$ is plotted against the reciprocal of substrate concentration $(1/[S])$ for experiments with the same initial enzyme concentration. The $y$-axis intercept of the plot is $1/v_{max}$ and the gradient is $K_M/v_{max}$.

**Related topics**

Rate law determination (F2)            Rate laws in action (F5)
Formulation of rate laws (F4)

---

**Chain reactions**

In many **complex reaction** systems, the product of one **elementary reaction** step is the reactant in the next elementary reaction step, and so on. Such systems are called **chain reactions** and the reactive intermediates responsible for the propagation of the reaction are called **chain carriers**. Important examples of such processes include combustion reactions in flames, the reactions that contribute to ozone destruction in the upper atmosphere, nuclear fission, or the formation of polymers in solution.

A chain reaction mechanism is illustrated by the gas-phase reaction between hydrogen and bromine:

$$H_2(g) + Br_2(g) \rightarrow 2HBr(g)$$

The following reaction scheme explains the complex observed **empirical rate law** (see Topic F2),

$$\text{rate of formation of HBr} = \frac{k[\text{H}_2][\text{Br}_2]^{\frac{3}{2}}}{[\text{Br}_2]+k'[\text{HBr}]}$$

1) **Initiation**. The initiation step is the **unimolecular** dissociation of $\text{Br}_2$ to produce the first **free radical** chain carriers. (Free radicals are reactive species containing unpaired electrons in their valence shells.)

$$\text{Br}_2 \rightarrow 2\text{Br} \qquad \text{rate} = k_1[\text{Br}_2]$$

2) **Propagation**. Propagation reactions convert reactive intermediates from a preceding elementary reaction into another reactive intermediate. The total number of reactive intermediates is unaltered. There are two different propagation reactions in the HBr mechanism:

$$\text{Br} + \text{H}_2 \rightarrow \text{HBr} + \text{H} \qquad \text{rate} = k_{2a}[\text{Br}][\text{H}_2]$$

and

$$\text{H} + \text{Br}_2 \rightarrow \text{HBr} + \text{Br} \qquad \text{rate} = k_{2b}[\text{H}][\text{Br}_2]$$

Although not present in the HBr mechanism, **branching** reactions are a specific type of propagation reaction in which more chain carriers are produced than are consumed.

3) **Retardation**. The attack of an H radical on a product HBr molecule formed in a previous propagation step, although still generating another free radical, has the effect of decreasing, or retarding, the overall rate of product formation.

$$\text{H} + \text{HBr} \rightarrow \text{H}_2 + \text{Br} \qquad \text{rate} = k_3[\text{H}][\text{HBr}]$$

4) **Termination**. Elementary reactions in which radicals combine to reduce the total number of radicals present are called termination steps.

$$\text{Br} + \text{Br} + \text{M} \rightarrow \text{Br}_2 + \text{M} \qquad \text{rate} = k_4[\text{Br}]^2$$

In this association reaction, the third body M represents any species present which removes the energy of the recombination collision between the Br atoms to form the stabilized $\text{Br}_2$ molecule. The concentration of M (which is a constant for given reaction conditions) is included in the value of the rate constant $k_4$. Although other chain termination reactions are possible, e.g. recombination of two H radicals, it turns out that only Br recombination is significant in this mechanism.

The rate law is formulated using the procedures described in Topic F4. Product HBr is formed in the two propagation reactions but consumed in the retardation reaction, so:

$$\text{rate of formation of HBr} = k_{2a}[\text{Br}][\text{H}_2] + k_{2b}[\text{H}][\text{Br}_2] - k_3[\text{H}][\text{HBr}]$$

Expressions for the concentrations of the Br and H intermediates are obtained by using the **steady state assumption**:

$$\text{rate of formation of Br}$$
$$= 2k_1[\text{Br}_2] - k_{2a}[\text{Br}][\text{H}_2] + k_{2b}[\text{H}][\text{Br}_2] + k_3[\text{H}][\text{HBr}] - 2k_4[\text{Br}]^2 = 0$$

and,

$$\text{rate of formation of H} = k_{2a}[\text{Br}][\text{H}_2] - k_{2b}[\text{H}][\text{Br}_2] - k_3[\text{H}][\text{HBr}] = 0$$

Solving the two equations gives the steady state concentrations:

$$[Br] = \left(\frac{k_1[Br_2]}{k_4}\right)^{\!\!1/2}$$

and

$$[H] = \frac{k_{2a}\left(\dfrac{k_1}{k_4}\right)^{\!\!1/2}[H_2][Br_2]^{1/2}}{k_{2b}[Br_2] + k_3[HBr]}$$

These are substituted into the rate expression for formation of HBr to obtain, after rearrangement:

$$\text{rate of formation of HBr } = \frac{2k_{2a}\left(\dfrac{k_1}{k_4}\right)^{\!\!1/2}[H_2][Br_2]^{3/2}}{[Br_2] + \left(\dfrac{k_3}{k_{2b}}\right)[HBr]}$$

This equation is the same as the empirical rate law with the empirical rate coefficients identified as:

$$k = 2k_{2a}\left(\frac{k_1}{k_4}\right)^{\!\!1/2} \text{ and } k' = \frac{k_3}{k_{2b}}$$

The agreement of this rate law with the empirical rate law does not prove that the proposed mechanism is correct but provides consistent evidence that it is correct. Further evidence could be obtained by laboratory measurement of values for the elementary rate coefficients (see Topic F1) and showing that the appropriate combinations correctly matched the values of the observed composite rate coefficients.

**Explosions**

**Chain reactions** which contain **chain branching** steps, i.e. reactions which increase the total number of **chain carriers**, have the potential for runaway reaction propagation and, under the right conditions, for **explosion**. Familiar examples are $H_2/O_2$ gas mixtures, or the hydrocarbon/$O_2$ mixtures that provide the explosive power in the cylinders of a car engine.

The $H_2/O_2$ reaction can be initiated in a number of ways, one of which is **bimolecular** collision between the two species to produce an H atom radical:

$$H_2 + O_2 \rightarrow H + HO_2$$

The H atom instigates a series of propagation and branching reactions so that after just a few reaction steps the number of H atoms has trebled:

|  |  |  |
|---|---|---|
| $H + O_2$ | $\rightarrow OH + O$ | Propagation and branching |
| $OH + H_2$ | $\rightarrow H + H_2O$ | Propagation |
| $O + H_2$ | $\rightarrow OH + H$ | Propagation and branching |
| $OH + H_2$ | $\rightarrow H + H_2O$ | Propagation |
| Net: $H + O_2 + 3H_2$ | $\rightarrow 3H + 2H_2O$ | |

This reaction scheme illustrates the ability of branching reactions to create extremely rapid growth in the number of chain carriers and the number of parallel elementary reactions. Whether or not a chain reaction ultimately leads to explosion depends on a number of factors such as the ratio of **chain termination** to chain branching processes, the initial concentration of reactants (which is a function of pressure for gas reactants such as $H_2$ and $O_2$), the temperature, and the rate at which energy (principally heat) can dissipate from the system. The complex dependence of $H_2/O_2$ explosion on pressure and temperature is shown in *Fig. 1*. The presence of a complex boundary between steady reaction and explosion reflects competition between the rates of different temperature and pressure dependent reactions in the mechanism. The system is difficult to interpret analytically because the **steady state approximation** (in which concentrations of reaction intermediates are assumed to remain constant) is not valid under the non-linear conditions of chain branching.

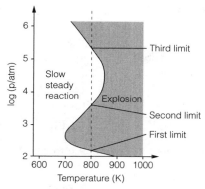

Fig. 1.    The boundary between steady reaction and explosion for the $H_2/O_2$ reaction as a function of pressure and temperature. The dashed line indicates the transitions between steady reaction and explosion as the pressure is increased for an example temperature of 800 K.

**Enzyme kinetics: the Michaelis-Menten equation**

A vast number of reactions in living systems are **catalyzed** by protein molecules called **enzymes** (see Topic F3). The lock and key hypothesis of enzyme action (*Fig. 2*) supposes that the enzyme, E, contains a very specific binding site into which fits only the target substrate, S, to form an enzyme-substrate complex, ES, which may undergo unimolecular decomposition back to E and S, or unimolecular reaction to form product, P, and the release of E for further reaction.

$$E + S \underset{k_{-1}}{\overset{k_1}{\rightleftharpoons}} ES \overset{k_2}{\longrightarrow} E + P$$

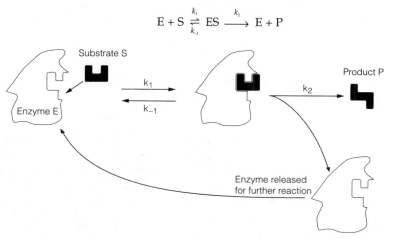

Fig. 2.    The lock and key mechanism for an enzyme catalyzed reaction.

The general kinetic mechanism parallels that of the **Lindemann mechanism** underpinning unimolecular reactions.

The overall rate of product formation is equal to $k_2[ES]$ and the value of $[ES]$ is evaluated by equating the rates of ES formation and removal under the **steady state** assumption:

$$k_1[E][S] = k_{-1}[ES] + k_2[ES]$$

or,

$$[ES] = \frac{k_1[E][S]}{k_{-1} + k_2}$$

If $[E]_0$ is the total concentration of enzyme present then $[E]_0 = [E] + [ES]$, and substituting for $[E]$ in the expression for $[ES]$ gives:

$$[ES] = \frac{k_1([E]_0 - [ES])[S]}{k_{-1} + k_2}$$

which rearranges to:

$$[ES] = \frac{k_1[E]_0[S]}{k_{-1} + k_2 + k_1[S]}$$

In enzyme kinetics, the symbol $v$ is often used to denote observed **reaction rate** so:

$$v = k_2[ES] = \frac{k_1 k_2[E]_0[S]}{k_{-1} + k_2 + k_1[S]}$$

By dividing both the numerator and denominator by $k_1$ and defining $(k_{-1}+k_2)/k_1$ as $K_M$, the **Michaelis constant**, the observed rate of reaction simplifies to:

$$v = \frac{k_2[E]_0[S]}{K_M + [S]}$$

This **rate law** is known as the **Michaelis-Menten equation** and shows that the rate of enzyme-mediated reaction is **first order** with respect to enzyme concentration. The overall rate depends on the concentration of substrate. At low substrate concentrations, $[S] \ll K_M$, so the rate of reaction is first order in substrate concentration as well as enzyme concentration:

$$v = \frac{k_2}{K_M}[E]_0[S]$$

When the substrate concentration is sufficiently high that $[S] \gg K_M$, the overall rate of reaction is **zero order** in $[S]$:

$$v = k_2[E]_0$$

The rate of reaction is independent of substrate concentration under these conditions because at any given time all active sites of the enzymes are filled and increasing the amount of substrate cannot increase the yield of product. The **rate determining step** is therefore the rate at which the ES complex reacts to form products. These conditions also correspond to the maximum rate of reaction:

$$v_{max} = k_2[E]_0$$

and $k_2$ is often termed the **maximum turnover number**.

**Enzyme kinetics: Lineweaver-Burke plots**

The **Michaelis-Menten equation** can be expressed in a different form by taking the reciprocal of both sides:

$$\frac{1}{v} = \frac{1}{k_2[E]_0} + \frac{K_M}{k_2[E]_0[S]}$$

Substituting for the maximum rate of reaction, $v_{max} = k_2[E]_0$, gives

$$\frac{1}{v} = \frac{1}{v_{max}} + \frac{K_M}{v_{max}}\frac{1}{[S]}$$

Therefore, the reciprocal of reaction rate is directly proportional to the reciprocal of substrate concentration when the total concentration of enzyme is held constant. The graph of $1/v$ plotted against $1/[S]$ is a straight line and is known as a **Lineweaver-Burke plot** (*Fig. 3*). The gradient and $y$-axis intercept of the Lineweaver-Burke plot are equal to $K_M/v_{max}$ and $1/v_{max}$, respectively, from which the maximum rate of reaction and the Michaelis constant can be calculated.

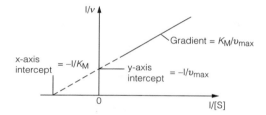

*Fig. 3.* *The Lineweaver-Burke plot for enzyme kinetics. The reciprocal of reaction rate, v, is plotted against the reciprocal of substrate concentrations, [S], for experiments using the same total enzyme concentration.*

# G1 NUCLEAR STRUCTURE

## Key Notes

**Nuclear structure**

Atomic nuclei are composed of protons and neutrons. A proton has +1 atomic charge and a neutron has zero charge. The atomic number, $Z$, of an atom is equal to the number of protons in the nucleus and is unique to each element. The atomic mass number, $A$, of an atom is equal to the sum of the number of protons and neutrons in the nucleus, so the number of neutrons, $N = A - Z$. A nucleus is represented by the chemical symbol for the element with a preceding subscript and superscript equal to $Z$ and $A$, respectively; for example, $^{12}_{6}C$.

**Isotope**

Nuclei which contain the same number of protons but different numbers of neutrons are called isotopes of that element. Isotopes undergo identical chemical reaction.

**Nuclear mass**

The unit of nuclear mass is the atomic mass unit, or amu, and is defined as exactly 1/12th the mass of a $^{12}_{6}C$ atom. The number of atoms in exactly 12.000 g of $^{12}_{6}C$ is called the Avogadro number and equals $6.0221 \times 10^{23}$.

**Binding energy**

The total nuclear mass of an atom is always slightly less than the sum of the individual masses of the constituent protons and neutrons. The difference is called the binding energy of the nucleus and arises because energy is released when neutrons and protons combine to form the nucleus. The relationship between binding energy and mass lost is $E = mc^2$.

**Relative atomic mass**

The relative atomic mass of an individual atom is the atomic mass relative to 1/12th the mass of an atom of carbon-12. The relative atomic mass of an element is the weighted average of the relative atomic masses of the naturally occurring isotopes.

**Nuclear stability**

One measure of nuclear stability is the binding energy per nucleon, which increases rapidly with atomic mass number, $A$, and reaches a maximum at $A \approx 56$. Nuclei with even numbers of protons and neutrons are more stable than nuclei with odd numbers of either or both. Nuclei with 2, 8, 20, 28, 50, 82 or 126 protons or neutrons are particularly stable.

**Related topics**

Applications of nuclear structure (G2)

Chemical and structural effects of quantization (G7)

---

**Nuclear structure**

The nucleus of an atom is composed of **protons** and **neutrons** bound together by short-range nuclear forces. The protons and neutrons are collectively known as **nucleons**. The charge on a proton is equal to +1 atomic charge unit and neutrons have zero charge. The number of protons in the nucleus is called the **atomic number**, or **atomic charge**, of the atom and has the symbol $Z$. The atomic number is unique to each element. In a neutral atom the positive charge of the protons in the nucleus is balanced by an equal number of negatively charged electrons in orbitals surrounding the nucleus.

The total number of protons and neutrons together in the nucleus is called the **atomic mass number** of the nucleus and has the symbol, $A$. The number of neutrons in the nucleus, $N$, is therefore $(A - Z)$. To represent a particular atomic nucleus, the chemical symbol is written with a preceding subscript equal to $Z$ and a preceding superscript equal to $A$. For example, an atom of carbon-12 (with 6 protons and 6 neutrons) is written as $^{12}_{6}C$. The $Z$ subscript is not strictly necessary since the chemical symbol also uniquely defines $Z$.

The nucleus constitutes only a very small fraction of the total volume of an atom. The length of nuclear radii lie in the approximate range of $(1 - 8) \times 10^{-15}$ m.

**Isotope**

The atomic number, or number of protons, $Z$, is unique to each element, but different atoms of a particular element may contain different numbers of neutrons and thus have different atomic mass numbers, $A$. Atoms that have the same number of protons but different numbers of neutrons are known as **isotopes** of that element. For example, oxygen has three isotopes whose nuclei contain either 8, 9 or 10 neutrons, in addition to the 8 protons. The isotopes are written as $^{16}_{8}O$, $^{17}_{8}O$ and $^{18}_{8}O$. Isotopes of an element undergo identical chemical reaction since the number of protons and electrons which define the bonding characteristics are the same for each isotope.

Most elements, like oxygen, have isotopes that are stable under normal conditions. Such **stable isotopes** do not emit ionizing radiation and do not spontaneously transform into atoms of a different element. Isotopes that are unstable and decompose naturally into other elements, with the release of harmful subatomic particles and/or radiation, are known as **radioisotopes** (Topic G2).

**Nuclear mass**

By definition, the **atomic mass unit**, or **amu** (also called Dalton, Da), is equal to 1/12th the mass of an atom of the $^{12}_{6}C$ **isotope** of carbon, i.e. the mass of $^{12}_{6}C$ equals exactly 12.000 amu. In practice this means that 1 amu = $1.660540 \times 10^{-27}$ kg. The number of atoms in exactly 12.000 g of $^{12}_{6}C$ atoms is known as the **Avogadro constant** or **Avogadro number, $N_A$**. Since the mass of one $^{12}_{6}C$ atom is equal to $12 \times 1.660540 \times 10^{-24}$ g, the value of the Avogadro number is:

$$N_A = \frac{12.000}{12 \times 1.660540 \times 10^{-24}} = 6.0221 \times 10^{23}$$

The Avogadro number of any entity is called a **mole** of that entity. Therefore, one mole of $^{12}_{6}C$ atoms contains $6.0221 \times 10^{23}$ atoms of carbon-12 and has a mass of 12.000 g.

The properties of some sub-atomic particles and stable nuclei are given in *Table 1*.

**Binding energy**

The total nuclear mass of an atom is always slightly less than the sum of the individual masses of the constituent protons and neutrons because energy is released when neutrons and protons combine to form the nucleus. The energy loss is called the **binding energy** of the nucleus. The magnitude of mass loss, $m$, is quantitatively related to the energy released, $E$, according to Einstein's equation:

$$E = mc^2$$

where $c$ is the velocity of light. Conversely, energy equal to, or greater than, the

Table 1.    Properties of some sub-atomic particles and stable nuclei

| Symbol | $Z$ | $A$ | Mass / amu | Isotopic abundance / % |
|---|---|---|---|---|
| $_{-1}^{0}\beta$ (electron) | −1 | 0 | $5.4858 \times 10^{-4}$ | – |
| $_{0}^{1}n$ (neutron) | 0 | 1 | 1.0086650 | – |
| $_{1}^{1}p$ (proton) | 1 | 1 | 1.0072765 | – |
| $_{1}^{1}H$ | 1 | 1 | 1.0078250 | 99.985 |
| $_{1}^{2}H$ (deuterium) | 1 | 2 | 2.0141018 | 0.015 |
| $_{2}^{4}He$ ($\alpha$-particle) | 2 | 4 | 4.0026033 | 100 |
| $_{6}^{12}C$ | 6 | 12 | 12 (exactly) | 98.90 |
| $_{6}^{13}C$ | 6 | 13 | 13.0033548 | 1.10 |
| $_{8}^{16}O$ | 8 | 16 | 15.9949146 | 99.762 |
| $_{8}^{17}O$ | 8 | 17 | 16.9991360 | 0.038 |
| $_{8}^{18}O$ | 8 | 18 | 17.9991594 | 0.200 |
| $_{17}^{35}Cl$ | 17 | 35 | 34.9688528 | 75.77 |
| $_{17}^{37}Cl$ | 17 | 37 | 36.9659026 | 24.23 |

binding energy must be supplied to separate a nucleus into its constituent nucleons.

**Relative atomic mass**

The **relative atomic mass** of an individual atom is the atomic mass relative to 1/12th the mass of an atom of carbon-12. By definition a relative atomic mass has no units. Since a natural sample of an element may contain a mixture of different **isotopes**, each having different atomic masses, the relative atomic mass of an element is equal to the weighted average of the naturally occurring isotopes. For example, chlorine has two naturally occurring isotopes, $_{17}^{35}Cl$ and $_{17}^{37}Cl$, with relative atomic masses of 34.96885 and 36.96590, respectively. (The relative atomic masses are not whole numbers because atomic masses of protons and neutrons are not whole numbers and mass is converted to **binding energy**.) In a natural sample of chlorine the $_{17}^{35}Cl$ and $_{17}^{37}Cl$ isotopes are present in the proportions 75.77% and 24.23%, respectively, so the overall relative atomic mass for natural chlorine is:

$$0.7577 \times 34.96885 + 0.2423 \times 36.96590 = 35.453$$

Therefore one **mole** (or **Avogadro's number**) of a natural sample of chlorine atoms has a mass of 35.453 g.

**Nuclear stability**

The **binding energy**, $E_b$, of a nucleus provides an indication of the total stability of the nucleus relative to the individual constituent nucleons. A more useful indicator of relative nuclear stability is the **binding energy per nucleon** which is the value of the binding energy of a particular nucleus divided by the total number of nucleons, $A$, in the nucleus. The value of $E_b/A$ as a function of $A$ is plotted in *Fig. 1*. After a sharp increase for the lightest elements the binding energy per nucleon remains fairly constant at around 8 MeV for $A \geq 16$ (elements heavier than O in the periodic table), which reflects the attainment of maximum packing around each individual nucleon once a minimum number have come together. The shallow maximum in $E_b/A$ for values of $A \approx 56$ (elements around Fe in the periodic table) indicates that these isotopes have enhanced relative stability. Because of this maximum, the **fission** (splitting) of a heavy nucleus into a pair of nuclei of approximate mass 56 is a process that releases energy. Similarly, the **fusion** (joining) of two of the lightest nuclei is also a process that releases energy.

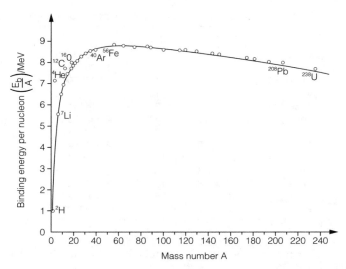

*Fig. 1.    Binding energy per nucleon as a function of mass number for some stable isotopes.*

Nuclei with even numbers of protons and neutrons are more stable than nuclei with odd numbers of either or both (*Table 2*). Only 8 nuclei are of the odd-odd type. In addition, nuclei with the 'magic' numbers of 2, 8, 20, 28, 50, 82, 126 protons or neutrons are particularly stable and abundant in nature. Such magic numbers arise from a shell model of the nucleus with an energy-level scheme analogous to the orbital energy-level scheme used for atomic electrons (see Topics G5 and G6).

*Table 2.    Frequency of occurrence of odd and even combinations of nucleons in stable nuclei*

| A | N | Z | Number of nuclei |
|---|---|---|---|
| Even | Even | Even | 166 |
|  | Odd | Odd | 8 |
| Odd | Even | Odd | 57 |
|  | Odd | Even | 53 |

# G2 APPLICATIONS OF NUCLEAR STRUCTURE

## Key Notes

**Radioactivity**

The isotopes of some elements are intrinsically unstable and will spontaneously disintegrate. Radioactivity is the emission of sub-atomic particles and/or electromagnetic radiation accompanying these conversions (transmutations) from parent isotope to daughter isotope. The three common forms of radioactive emission are: α-particles (helium-4 nuclei); β-particles (electrons); and γ-rays (very short-wave electromagnetic radiation). The time taken for 50% of a radioactive sample to decay is called the half-life. Radioactivity is measured in units of bequerel and the absorbed dose in units of gray.

**Isotope effects**

The kinetic isotope effect is the reduction in the rate of reaction by the replacement of an atom in a molecule by a heavier isotope (usually the replacement of hydrogen by deuterium). The effect is caused by the lowering of the zero-point energy of the X–H bond by the heavier atom which increases the activation energy required to break the bond. Observation of an isotope effect indicates that cleavage or formation of the bond forms part of the rate determining step.

**Isotope labeling**

Both radioactive and stable isotopes can be used to tag specific molecules to elucidate chemical reaction mechanisms. Molecules containing radioactive markers can be identified by the radiation emitted. Molecules enriched with a particular stable isotope are commonly identified using mass spectrometry.

**Related topic**

Nuclear structure (G1)

**Radioactivity**

The **isotopes** of a number of elements are naturally unstable and will lose mass and/or energy in order to form a more stable state. The spontaneous decay of such **radioisotopes** (or **radionuclides**) creates a different element (the **daughter**) from the starting element (the **parent**). The conversion of one isotope into another is called **transmutation**. Radioisotopes can also be prepared synthetically *via* the deliberate bombardment of stable nuclei with sub-atomic particles.

In all cases the mass or energy loss during radioactive decay can cause damage to the environment through the formation of reactive ions or free radicals. The extent of damage depends on the type of mass or energy emitted. In general, three types of ionizing radiation can be emitted during radioactive decay; **α-particles**, **β-particles** or **γ-radiation**.

The activity of a radioactive source is measured in **bequerel** (Bq) which is defined as one nuclear disintegration per second. (The older unit of **curie** (Ci) is equal to $3.7 \times 10^{10}$ disintegrations per second.) The unit of absorbed dose is the **gray** (Gy) which is defined as one joule per kilogram.

### α-decay

An α-**particle** consists of two protons and two neutrons and is effectively the nucleus of a helium-4 atom, $^4_2$He. Therefore, the daughter isotope resulting from emission of an α-particle has **atomic mass number**, $A$, four units less than the parent isotope and atomic number, $Z$, two units less, and is an isotope of the element two places to the left in the periodic table. Decay by α-particle emission usually only occurs amongst the heaviest elements which have mass numbers greater than 200. An example of α-decay is the transmutation of the uranium-238 isotope to an isotope of thorium:

$$^{238}_{92}U \rightarrow {}^{234}_{90}Th + {}^4_2He$$

α-particles are the most massive and highly charged (+2 charge units) of the particles emitted during spontaneous radioactive decay and travel relatively slowly, approximately 10% of the speed of light. Since α-particles readily lose their energy and neutralize their charge in collisions with surrounding elements, their effects are short-range (a few millimeters) and, in general, α-emitters are not considered particularly hazardous since they cannot penetrate through skin. However, they may cause burns to the outer layers of skin and are dangerous if ingested.

### β-decay

A β-**particle** is an electron. It therefore has a small mass, and a charge of –1 atomic charge unit. β-decay occurs when a neutron spontaneously converts into a proton, which remains in the nucleus, and an electron, which is emitted. Consequently, the parent and daughter nuclides have identical atomic mass number, $A$, but the daughter nuclide is an isotope of the element one atomic unit higher than the parent. An example is the β-decay of the carbon-14 radioisotope into the equivalent mass number isotope of nitrogen:

$$^{14}_6C \rightarrow {}^{14}_7N + {}^0_{-1}e$$

The velocities of β-particles are greater than those of α-particles, because of the much lighter mass, and it is possible for high-energy β-particles to penetrate skin and reach internal organs.

### γ-decay

**Gamma (γ) rays** are very high energy photons often released during α- and β-decay processes when the daughter nuclide decays from an excited to a more stable state. Since γ-rays are photons of the **electromagnetic spectrum** they have no mass and no charge and travel at the speed of light. Photons of γ-rays are even more energetic than those of X-rays and are thus extremely penetrating and highly damaging. The subsequent β-decay of the daughter thorium-234 isotope from the α-decay of uranium-238 is accompanied by emission of γ-rays:

$$^{234}_{90}Th \rightarrow {}^{234}_{91}Pa + {}^0_{-1}e + {}^0_0\gamma$$

### Half-life

It is not possible to predict exactly when an individual radioactive nucleus will spontaneously undergo radioactive decay. However, in a sample of such nuclei it is always the case that a fixed proportion of the sample will undergo radioactive decay within a fixed time-span. The time taken for half the sample to decay is known as the **half-life**. Since radioactive decay is a **first order** process the kinetics of the decay are described by **first order kinetics** (see Topic F2). The half-lives of different radioactive isotopes can vary between fractions of a second to billions of years. A selection of important radioisotopes with their decay processes and half-lives is given in *Table 1*.

*Table 1. Some important radioactive nuclei and their modes of decay*

| Element | Isotope | Mode of decay | Half-life |
|---|---|---|---|
| **Natural isotopes** | | | |
| Uranium | $^{238}_{92}$U | $\alpha$ | $4.5 \times 10^9$ years |
| Radium | $^{226}_{88}$Ra | $\alpha, \gamma$ | 1600 years |
| Radon | $^{222}_{86}$Rn | $\alpha$ | 3.8 days |
| Carbon | $^{14}_{6}$C | $\beta$ | 5730 years |
| Potassium | $^{40}_{19}$K | $\beta, \gamma$ | $1.3 \times 10^9$ years |
| **Synthetic isotopes** | | | |
| Hydrogen (tritium) | $^{3}_{1}$H | $\beta$ | 12.3 years |
| Phosphorus | $^{32}_{15}$P | $\beta$ | 14.3 days |
| Cobalt | $^{60}_{27}$Co | $\beta, \gamma$ | 5.27 years |
| Cesium | $^{137}_{55}$Cs | $\beta$ | 30.1 years |

**Isotope effects**

It is often observed that the **rate** of chemical reaction is reduced when an atom forming a bond that must be cleaved during the **rate determining step** of the reaction (see Topic F4) is replaced by a heavier **isotope** of the same element. This is the **kinetic isotope effect** and is particularly apparent when a hydrogen atom, $^1$H, is replaced by a deuterium atom, $^2$H, since this produces the greatest relative mass change of any isotopic substitution. The isotope effect arises because the **zero-point energy** of the X–H bond undergoing cleavage is lowered when H is replaced by D. (The substitution of a heavier isotope does not alter the strength, or force constant, of the bond but the heavier mass lowers the equilibrium oscillation frequency of the bond which reduces the separation of the vibrational energy levels, Topic I4.) This lowering of zero-point energy increases the magnitude of the **activation energy** of the reaction (*Fig. 1*) which reduces the rate of the reaction (see Topic F3). For example, in the oxidation reaction:

$$Ph_2CHOH \xrightarrow{MnO_4^-,\ OH^-} Ph_2C{=}O$$

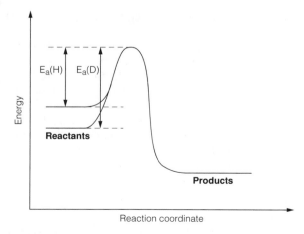

*Fig. 1. The origin of the kinetic isotope effect. Activation energy is increased when a hydrogen atom is replaced by the deuterium isotope.*

$Ph_2CHOH$ is oxidized 6.7 times as rapidly as $Ph_2CDOH$, from which it is deduced that scission of the C–H (or C–D) bond is involved in the rate determining step. In contrast, benzene, $C_6H_6$, and hexadeuterobenzene, $C_6D_6$, undergo nitration at essentially the same rate:

so C–H bond breaking, which must occur at some stage in the overall process, cannot be involved in the rate-determining step.

Effects arising from substitution of an atom directly constituting a bond that is broken or formed during the rate determining step are known as **primary kinetic isotope effects**, whereas **secondary kinetic isotope effects** arise from isotopic substitution elsewhere in the molecule.

**Isotope labeling**  A useful application of both **stable** and **radioactive isotopes** is as tracers to identify specific molecules in order to elucidate chemical or biochemical reaction mechanisms. For example, the radioactive isotope of phosphorus, $^{32}P$, is routinely used during the analysis of DNA or RNA molecules, both of which contain phosphate linkages within the polymer chain. The $^{32}P$ tag is introduced into the DNA or RNA molecules from donor $^{32}P$-labeled molecules using specific phosphoryl transfer catalyzing enzymes. The DNA or RNA molecules are then cleaved at specific sites using another enzyme and the resulting mixture of fragments separated along a polyacrylamide gel by **electrophoresis** (see Topic E8). The position of bands which contain the labeled DNA are determined by the darkening of a photographic plate from the electrons emitted during the β-decay. The sensitivity towards detection of radioactivity means that only very small amounts of radioactive material are required for analysis.

The detection of radioactive $^{14}C$, with half-life of 5370 years, is used extensively to trace metabolic pathways in cells. It is also used to determine the age of ancient natural materials. In this application the source of $^{14}C$ is the continual natural transmutation of nitrogen in the atmosphere by cosmic rays. The metabolic processes in living material maintain an equilibrium in the fraction of $^{14}C$ present within the carbon of the plant or animal. When the plant or animal dies there is no more active exchange with the source of $^{14}C$ and the fraction of $^{14}C$ remaining in the material decreases at a rate determined by its half-life.

Some elements do not have a radioactive isotope of convenient half-life and in these instances stable isotopes can be used in tracer experiments. For example, methanol which has been synthesized so that the oxygen atom is abnormally enriched with the $^{18}O$ isotope can be used to identify whether the starred oxygen atom in the following ester product comes originally from the acid or the alcohol starting reagents:

The isotopic composition of the product is characterized using a mass spectrometer, which is an instrument for measuring the mass of molecules. Enrichment of the ester with $^{18}O$ isotope proves that the oxygen atom in the ester linkage must come from the alcohol and not the acid.

# G3 QUANTIZATION OF ENERGY AND PARTICLE-WAVE DUALITY

## Key Notes

**The failures of classical physics**

Classical physics assumes that particles move along precisely defined trajectories and can possess any amount of energy. The failure of classical physics to account for observed phenomena such as black body emission and the photoelectric effect was resolved by the postulates of quantization and particle wave-duality and showed that classical mechanics was an approximate description of a more fundamental quantum mechanics.

**Quantization**

Quantization is the confinement of a property (such as energy, momentum or position in space) to a set of discrete values, called quanta.

**The Planck constant**

The Planck constant, $h$, is the constant of proportionality between a quantum of energy, $E$, and the frequency, $v$, of the corresponding photon of electromagnetic radiation, $E = hv$. Its value is $6.626 \times 10^{-34}$ J s.

**The ultraviolet catastrophe**

The classical physics interpretation of the power emitted by a black body assumes that electromagnetic oscillators can oscillate at all frequencies. This leads to the ultraviolet catastrophe in which black body emission is predicted to increase to infinity at high radiation frequency. The postulates that energy is quantized according to frequency, and that oscillators can only be excited by energy equal to the quanta, resolves the problem.

**The photoelectric effect**

The photoelectric effect is the emission of electrons from a surface irradiated by ultraviolet light. No electrons are emitted unless the radiation frequency exceeds a threshold value characteristic of the surface. The kinetic energy of the electrons varies linearly with the frequency of the radiation and is independent of the intensity of the radiation. The effect is evidence that radiation is quantized into particles (photons) with energy proportional to frequency.

**Young's slit experiment**

Light passing through two closely spaced narrow slits produces a diffraction pattern of alternating dark and light fringes, readily interpreted in terms of constructive and destructive interference of wave fronts passing through the slits. The observation that particles produce the same effect is evidence of particle-wave duality.

**De Broglie's equation**

The de Broglie equation summarizes the relationship between particle momentum, $p$ ($= mv$) and wavelength, $\lambda$, in the particle-wave duality interpretation of matter and radiation, $p = h/\lambda$.

**Related topics**          Diffraction by solids (A6)          The wave nature of matter (G4)

**The failures of classical physics**

In the everyday world of macroscopic objects the Newtonian laws of **classical physics** account extremely well for the motion of particles along defined trajectories. These laws assume that the position and velocity of a particle can be defined at every instant, from which it is possible, at least in theory, to calculate the precise position and velocity of the particle at every other instant. Classical physics further assumes that any type of motion can be supplied with any arbitrary amount of energy. Thus, for example, the range of an artillery shell is, in principle, continuously variable according to the amount of energy supplied at the initial firing.

However, it turns out that the laws of classical mechanics are an approximate description of the motion of particles, accurate only in the limit of large objects travelling at velocities much less than the speed of light. To account for the behavior of very small particles such as molecules, atoms or electrons requires the application of a more fundamental set of laws, the laws of **quantum mechanics**. Quantum theory was formulated in the early years of the 20th century when classical physics failed to account for many sets of experimental observations arising from atomic-scale phenomena, for example, the **ultraviolet catastrophe** and the **photoelectric effect**. The resolution of these failures incorporated the postulate that energy was **quantized** and introduced the concept of **particle-wave duality** for radiation and matter.

**Quantization**

A fundamental outcome of the theory of quantum mechanics is that properties such as energy are no longer permitted to assume any value within a continuum but are confined to a series of discrete values only. This outcome is called **quantization** and the discrete values are called **quanta**. The values of the quanta depend on the specific **boundary conditions** of the system under consideration (Topic G4). Other properties to which quantization applies include position and angular momentum.

**The Planck constant**

The **Planck constant**, $h$, is a fundamental constant of quantum theory and appears in very many equations describing quantum mechanical phenomena. It is the constant of proportionality between the energy, $E$, of a **photon** and the frequency, $v$, of the associated **electromagnetic radiation**:

$$E = hv$$

Planck's constant can be determined from an analysis of the **photoelectric effect**. The value is $6.626 \times 10^{-34}$ J s. So, for example, an ultraviolet photon of wavelength 300 nm (frequency, $v = c/\lambda = 9.993 \times 10^{14}$ Hz) has an energy of $6.62 \times 10^{-19}$ J. One **mole** of these photons has an energy $6.022 \times 10^{23} \times 6.62 \times 10^{-19} = 399$ kJ which exceeds the energy of many chemical bonds and explains why ultraviolet radiation can damage molecules in materials and biological cells.

**The ultraviolet catastrophe**

The attempt to derive an expression for the power emitted by a **black body** as a function of wavelength was an early example of the failure of classical physics. A black body is a perfect emitter and absorber of electromagnetic radiation, capable of emitting and absorbing all frequencies of radiation uniformly. A good approximation of a black body is a pinhole in a container maintained at a uniform temperature.

The graph of the observed spectral output of a black body (the power emitted as a function of wavelength) has a characteristic shape in which the power increases through a maximum as the wavelength decreases (*Fig. 1*). At higher temperatures $T$ the wavelength of the peak emission shifts to shorter wavelengths, and the total power emitted (the area under the curves in *Fig. 1*) increases proportionally as $T^4$. For example, the embers of coal in a fire glow red whereas the color of a much hotter object, such as the surface of the sun, appears yellow-white because of the greater contribution from shorter wavelength blue light to the visible part of its emission.

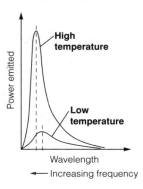

Fig. 1.  Power emitted as a function of wavelength for two temperatures of a black body.

In attempting to derive a formula to account for the shape of the curves in *Fig. 1* the physicists Rayleigh and Jeans assumed that the oscillators which comprise electromagnetic radiation could oscillate at any frequency and therefore that all wavelengths, $\lambda$, of radiation were possible. The resulting equation:

$$\text{power emitted in wavelength range } \lambda + \delta\lambda = \frac{8\pi kT}{\lambda^4}\delta\lambda$$

works fairly well at long wavelengths (low frequencies) but fails at short wavelengths (high frequencies) because as $\lambda$ decreases the power emitted increases continuously towards infinity, and never passes through a maximum. The equation predicts that a black body is a strong emitter of all wavelengths, including ultraviolet, X-rays and $\gamma$-rays, even at room temperature. This obvious absurdity is termed the **ultraviolet catastrophe**.

The problem is resolved by Planck's postulate that the energy of each electromagnetic oscillator is limited to discrete values of energy equal to an integral multiple of its oscillation frequency, $v$:

$$E = nhv \qquad n = 0, 1, 2...$$

The constant of proportionality, $h$, is **Planck's constant** ($6.626 \times 10^{-34}$ J s). The consequence of this **quantization** is that oscillators can only be stimulated when energy of value $hv$ (or $2hv$, or $3hv$, etc.) is available. The relative probability of finding oscillators of energy $nhv$ at a temperature $T$ is given by the $e^{-nhv/k_BT}$ factor of the **Boltzmann distribution** law (see Topic G8). This factor tends to zero as the value of $v/T$ in the exponential increases. Since the values of $hv$ for X-rays or $\gamma$-rays are very large (very high frequencies of oscillation) only a negligible fraction of these oscillators are stimulated unless the energy (or $T$) of the black body is itself extremely large. Planck's modified version of the energy density formula for a black body includes the Boltzmann exponential term:

$$\text{power emitted in wavelength range } \lambda + \delta\lambda = \frac{8\pi hc}{\lambda^5}\left(\frac{1}{e^{\frac{hc}{\lambda k_B T}} - 1}\right)\delta\lambda$$

and reproduces the experimental curve in *Fig. 1* extremely well. At large values of $\lambda$ the Planck law and Rayleigh-Jeans law are equivalent.

**The photoelectric effect**

The **photoelectric effect** is the emission of electrons from a surface (usually a metal) when the surface is irradiated with ultraviolet light. The maximum kinetic energy of the ejected electrons, $1/2m_e v^2$, can be calculated from the threshold negative voltage required to repel them from a detector above the surface. Three key experimental observations of the photoelectric effect are:

(i)   no electrons are ejected, regardless of the intensity of the radiation, unless the frequency of the radiation exceeds a threshold value characteristic of the metal;

(ii)  once the threshold frequency is exceeded the kinetic energy of the ejected electrons is linearly proportional to the frequency of the incident radiation;

(iii) the kinetic energy of the ejected electrons does not depend on the intensity of the incident radiation. Only the number of ejected electrons depends on the intensity.

These observations cannot be reconciled with the classical interpretation of electromagnetic waves in which the energy of a wave is assumed to depend on its amplitude and not on its frequency.

As with the **ultraviolet catastrophe**, the photoelectric effect is explained by the postulate that the energy of the incoming electromagnetic radiation is **quantized** such that radiation of frequency $v$ consists only of quanta of energy, $E = hv$. Therefore the energy of the radiation depends only on the value of $v$; increasing the intensity of radiation at this frequency increases the number of quanta present ($n = E/hv$) but does not change the energy of each quantum. The quanta of electromagnetic radiation are called **photons**. If it is assumed that each metal has a characteristic energy barrier for ejection of an electron (called the **work function, $\Phi$**) then only radiation with quanta of energy that exceed $\Phi$ can liberate photoelectrons (*Fig. 2*). Above the threshold frequency the kinetic

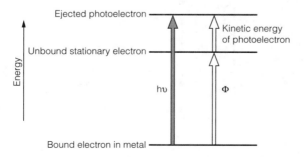

Fig. 2.   *Energetics of the photoelectric effect. The difference in energy between the incoming photon and the work function of the metal appears as kinetic energy in the ejected photoelectron.*

energy of the photoelectrons increases linearly with the energy difference between the incoming photons and the work function:

$$\frac{1}{2}m_e v^2 = hv - \Phi$$

The equation can be used to derive the value of **Planck's constant**. A graph of photoelectron kinetic energy against frequency of radiation is a straight line with gradient $h$ and negative $y$-intercept equal to the magnitude of the work function for that surface (*Fig. 3*).

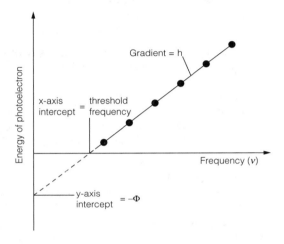

Fig. 3.    *Plot of photoelectron energy vs. irradiation frequency. The gradient is Planck's constant.*

**Young's slit experiment**

The Young's slit experiment is the production of a distinctive **diffraction pattern** of alternating dark and light fringes when light from a single source passes through two closely spaced narrow slits (*Fig. 4*). The effect is readily explained in the wave interpretation of light as arising from the alternating constructive interference (bright areas) and destructive interference (dark areas) of wave fronts emanating from the two slits.

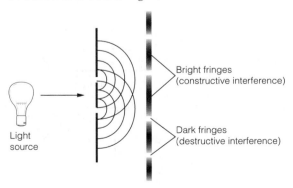

Fig. 4.    *Young's slit experiment. A light source directed at two slits produces a diffraction of pattern of alternating bright and dark stripes.*

The importance of the experiment in terms of quantum theory arises from the further observation that it is possible to acquire a diffraction pattern even when the intensity of the light source is reduced to sufficiently low levels that calculations show there can only be one photon at a time between source and screen. Thus wave-like properties remain even at the limit of a single photon. A diffraction pattern can also be obtained using a beam of particles, e.g. electrons and neutrons, rather than light as the source (see Topic A6).

**De Broglie's equation**

Experimental observations such as the **photoelectric effect**, the **Young's slit experiment** and **diffraction** by electrons show that, at the atomic scale, it is not possible to think of radiation and matter as consisting uniquely of either waves or particles. Instead both radiation and matter jointly exhibit properties associated with both waves and particles, and either interpretation is equally valid. This joint particle-wave character of matter and radiation is known as **particle-wave duality**.

The quantitative link between the wavelength of a particle and its linear momentum $p$ (the product of the particle's mass and velocity, $p = mv$) is the **de Broglie equation**:

$$\lambda = \frac{h}{p}$$

where $h$ is **Planck's constant**. The de Broglie relationship implies that faster moving particles have shorter wavelengths and that, for a given velocity, heavier particles have shorter wavelengths than lighter ones.

The relationship is confirmed by electron diffraction experiments in which the wavelength of the associated diffraction pattern matches the calculated momentum of the electrons. For example, the wavelength of an electron accelerated from rest by a voltage of 2.0 kV is calculated as follows. The kinetic energy, $\frac{1}{2}m_ev^2$, attained by the electron is equal to $eV$, where $V$ is the acceleration voltage and $e$ the charge on the electron ($1.602 \times 10^{-19}$ C). Since momentum, $p = mv$:

$$eV = \frac{1}{2}m_ev^2 = \frac{1}{2}\frac{p^2}{m_e}$$

Combining with de Broglie's equation gives:

$$\lambda = \frac{h}{\sqrt{2m_eeV}}$$

and a numerical value of $\lambda = 2.7 \times 10^{-11}$ m in this example.

A wavelength of 27 pm is comparable to the spacing between molecules in crystalline solids and explains how a beam of electrons can produce a diffraction pattern from a crystal. For comparison, the wavelength of a cricket ball of mass 0.1 kg travelling at 10 m s$^{-1}$ is $6.63 \times 10^{-34}$ m. This wavelength is many orders of magnitude smaller than a sub-atomic particle and shows why quantum mechanical phenomena are not important for macroscale objects.

# G4 THE WAVE NATURE OF MATTER

## Key Notes

**Wavefunctions and probabilities**

A wavefunction describes the region of space in which the particle it represents is located. The square of the wavefunction is proportional to the probability of finding the particle at that location.

**Schrödinger equation**

The Schrödinger equation, $H\psi_i = E_i\psi_i$ is the basic equation for calculating the wavefunctions $\psi_i$ of a quantum mechanical system described by a Hamiltonian operator $H$ (usually the sum of the kinetic and potential energy). Each wavefunction is associated with a specific energy $E_i$ of the system.

**Boundary conditions**

The imposition of boundary conditions on the solutions of the Schrödinger equation restricts a system to a set of physically allowable wavefunctions (and energies) and is the origin of quantization. An example boundary condition requires the value of the wavefunction to be zero at the wall of an infinitely deep potential well.

**Heisenberg's uncertainty principle**

The magnitude of the uncertainty which must co-exist between the position and momentum of a particle is $\Delta p \Delta x \geqslant \hbar/2$.

**Particle in a box**

Several general features of quantum mechanical systems are illustrated by the solution of the Schrödinger equation for a particle (mass $m$) constrained to one-dimensional motion between walls of infinite potential a distance $L$ apart (the box): the energy is quantized, $E_n = n^2h^2/8mL^2$; energy levels are more closely spaced for a larger box; the probability of finding the particle at different positions within the box is not uniform; the system possesses intrinsic zero point energy.

**Zero point energy**

The zero point energy is the minimum energy a system can possess. It is frequently non-zero as a consequence of Heisenberg's uncertainty principle.

**Particle in a circular orbit**

The energy of a particle undergoing rotational motion with moment of inertia $I = mr^2$ is quantized, $E_n = n^2\hbar^2/2I$. Both positive and negative values of $n$ are allowed because the particle can rotate with the same energy in either direction.

**Degeneracy**

Two or more states of a system are degenerate if they possess the same energy.

**Quantum tunneling**

Quantum tunneling is the probability of observing a particle beyond a (non-infinite) potential energy barrier that exceeds the energy of the particle. The effect arises because the amplitude of a wavefunction decreases exponentially within the barrier resulting in non-zero wavefunction amplitude beyond the barrier.

**Related topics**　　Quantization of energy and　　　Many-electron atoms (G6)
　　　　　　　　　　　particle-wave duality (G3)　　General features of spectroscopy
　　　　　　　　　　The structure of the hydrogen　　　(I1)
　　　　　　　　　　　atom (G5)

**Wavefunctions and probabilities**

In the **particle-wave duality** interpretation of matter and radiation (Topic G3) a particle moving in space can also be described as a wave in space with a wavelength related to the particle momentum by **de Broglie's equation**, $\lambda = h/p$. In **quantum mechanics**, the notion of a particle moving in defined trajectories in a system is replaced entirely by this description of the system in terms of its **wavefunctions, $\psi$**. The wavefunction simultaneously describes all regions of space in which the particle it represents can be found. This, in turn, introduces the idea of **uncertainty** into quantum mechanics because the exact position of the particle at each point in time is not defined, only the region of space of all its possible positions. The exact shape of the wavefunction is important because the **probability** of finding the particle at each point is proportional to $\psi^2$ at that point; a greater amplitude in the wavefunction corresponds to a greater probability density in the particle's distribution.

**Schrödinger equation**

The **Schrödinger equation** is the fundamental equation of quantum mechanics and has the general form:

$$H\psi_i = E_i\psi_i$$

Each allowed **wavefunction** $\psi_1$, $\psi_2$, $\psi_3$... of a system described by a **Hamiltonian operator**, $H$, is associated with one particular allowed **energy level** $E_1$, $E_2$, $E_3$... (An operator is a mathematical function that represents the action of a physical observable.) The Hamiltonian operator is the operator for the total kinetic and potential energy of the system. Only an allowed wavefunction of the system, $\psi$, when operated on by $H$, returns the same wavefunction, multiplied by the associated constant value $E$. In mathematical terminology the Schrödinger equation is an **eigenvalue** equation; the pairs of $E$ and $\psi$ that satisfy the equation are the **eigenvalues** and **eigenfunctions** of $H$, respectively.

A Schrödinger equation can be written to describe any particular physical system. For a particle of mass $m$ moving in one dimension only (along the $x$-axis) the equation is:

$$-\frac{\hbar^2}{2m}\frac{d^2\psi}{dx^2} + V(x)\psi = E\psi$$

where $-\hbar^2/2m\ d^2/dx^2$ and $V(x)$ are the operators for (one-dimensional) kinetic and potential energy, respectively, that together constitute the Hamiltonian operator. The symbol $\hbar$ is short-hand notation for $h/2\pi$ ($h$ is **Planck's constant**).

The Schrödinger equation can be shown to be consistent with experimental observation by considering the equation for a freely-moving particle that possesses kinetic energy only:

$$-\frac{\hbar^2}{2m}\frac{d^2\psi}{dx^2} = E\psi$$

Rearranging gives:

$$\frac{d^2\psi}{dx^2} = -\frac{2mE}{\hbar^2}\psi$$

A solution to this equation is:

$$\psi = \sin\sqrt{\frac{2mE}{\hbar^2}}x$$

which may be verified by differentiating the function twice:

$$\frac{d\psi}{dx} = \sqrt{\frac{2mE}{\hbar^2}}\cos\sqrt{\frac{2mE}{\hbar^2}}x$$

$$\frac{d^2\psi}{dx^2} = -\frac{2mE}{\hbar^2}\sin\sqrt{\frac{2mE}{\hbar^2}}x = -\frac{2mE}{\hbar^2}\psi$$

The wavelength, $\lambda$, of a sine wave of form $\sin(kx)$ is:

$$\lambda = \frac{2\pi}{k}$$

so the wavelength of the wavefunction associated with the freely-moving particle is:

$$\lambda = 2\pi\sqrt{\frac{\hbar^2}{2mE}}$$

Substituting using the relationship between kinetic energy $E$ and momentum $p$:

$$E = \frac{1}{2}mv^2 = \frac{(mv)^2}{2m} = \frac{p^2}{2m}$$

and remembering that $\hbar = h/2\pi$ gives:

$$\lambda = 2\pi\sqrt{\frac{h^2}{4\pi^2 2m}\frac{2m}{p^2}} = \frac{h}{p}$$

The final result is **de Broglie's equation**, i.e. the Schrödinger equation reproduces the experimental observation that a freely moving particle can be described as a sine wave of wavelength inversely proportional to the particle momentum.

**Boundary conditions**

In principle, there are an infinite number of solutions to the **Schrödinger equation**. If $\sin(kx)$ is a solution then so is $a\sin(bkx)$ for all values of $a$ and $b$. However, only a sub-set of solutions are allowed physically and these are determined by the **boundary conditions** imposed by the physical situation which the Schrödinger equation describes. Examples are shown for a **particle in a box** or an electron in the **hydrogen atom** (Topic G5). The fact that only certain values of $E$ and $\psi$ are allowed solutions of a particular Schrödinger equation is the origin of the **quantization** of energy (Topic G3).

**Heisenberg's uncertainty principle**

The **wavefunction** description of a moving particle replaces the classical concept that the particle moves with known velocity along a precisely defined trajectory. Combination of the **Schrödinger** and **de Broglie equations** shows that the wavefunction of a particle of momentum $p$ moving freely in the $x$ direction is:

$$\psi = \sin\frac{2\pi}{\lambda}x$$

with wavelength $\lambda = h/p$. The wavefunction has constant wavelength and peak-to-peak amplitude at all positions, corresponding to equal probability of finding the particle at any of an infinite number of points in the $x$ direction. Therefore, although the momentum of the particle is known exactly, its position is uncertain. The converse situation, in which the position of the particle in space is known exactly, requires a wavefunction which has zero amplitude everywhere except at the particle's position (*Fig. 1a*), and can only be achieved through the superposition of an infinite number of wavefunctions of different wavelengths, corresponding to an infinite range in particle momentum (*Fig. 1b*). These outcomes are encapsulated in **Heisenberg's Uncertainty Principle**:

*It is impossible to specify simultaneously both the position and momentum of a particle exactly.*

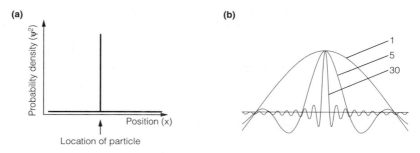

Fig. 1. ( a) The wavefunction of a particle with a well-defined position. (b) The superposition of a number of wavefunctions of different wavelengths. The superposition of an infinite number of wavefunctions of different wavelength is required to produce the spike wavefunction of a particle with a well-defined position.

The magnitude of the uncertainty which must coexist between position and momentum is quantitatively given by:

$$\Delta p \Delta x \geq \frac{\hbar}{2}$$

where $\Delta p$ and $\Delta x$ are the uncertainties in momentum and position, respectively. The value of $\hbar/2$ is very small so the phenomenon is not directly observable at the scale of everyday macroscopic objects. For example, the uncertainty in position of an object of mass 1.0 kg travelling with a velocity known to be better than $1.0 \times 10^{-3}$ m s$^{-1}$ precision is $5.3 \times 10^{-26}$ m. This uncertainty is many orders of magnitude smaller than the size of an atomic nucleus. However, the same uncertainty in velocity for an electron of mass $9.11 \times 10^{-31}$ kg implies an uncertainty in electron position far larger than the size of an atom.

**Particle in a box**

The application of **Schrödinger's equation** to a particle undergoing one-dimensional translational motion between confined limits demonstrates how imposition of **boundary conditions** gives rise to one of the fundamental principles of quantum mechanics, **quantization**. The two walls of the box are at positions $x = 0$ and $x = L$ along the $x$-axis. Inside the box the particle (mass $m$) moves freely in the $x$-direction, and the potential energy $V = 0$. The potential energy rises abruptly to infinity at the walls.

The Schrödinger equation for the particle in the box is:

$$-\frac{\hbar^2}{2m}\frac{d^2\psi}{dx^2} = E\psi$$

A solution to this equation is (see section on Schrödinger equation):

$$\psi = \sin\sqrt{\frac{2mE}{\hbar^2}}x$$

In fact, the general solution is:

$$\psi = a\,\sin\sqrt{\frac{2mE}{\hbar^2}}x$$

where any value of $E$ and $a$ forms a suitable wavefunction. However, because the particle is confined to a box of finite length, the walls impose **boundary conditions** on which wavefunctions are physically allowable. Since the potential energy rises to infinity at the walls the probability of finding the particle outside the box is zero, so the wavefunction at the walls of the box, and everywhere outside the box, must be zero. Therefore, all acceptable **wavefunctions** for the particle must fit exactly inside the box, like the vibrations of a string fixed at both ends. To satisfy this condition requires that the wavelength, $\lambda$, of allowed wavefunctions must be one of the values:

$$\lambda = 2L, L, \frac{2L}{3}, \frac{2L}{4}, \ldots$$

or, more concisely, that:

$$\lambda = \frac{2L}{n} \qquad n = 1, 2, 3, \ldots$$

The relationship between $\lambda$ and the mathematical description of a sine wave is $\sin(2\pi x/\lambda)$ so the wavelength of the wavefunction $\psi = a\,\sin\sqrt{\frac{2mE}{\hbar^2}}x$ is:

$$\lambda = \frac{1}{2\pi}\sqrt{\frac{\hbar^2}{2mE}}$$

Therefore, allowed wavefunctions of the particle in a box must satisfy:

$$\frac{2L}{n} = \frac{1}{2\pi}\sqrt{\frac{\hbar^2}{2mE}}$$

which, on rearranging gives:

$$E_n = \frac{n^2h^2}{8mL^2}$$

The subscript $n$ is added to emphasize the fact that the particle in the box is only permitted to possess discreet values, or **quanta**, of $E$, corresponding to integer values of $n$ in the above expression. The allowed energies of a system are called the **energy levels**. The wavefunctions associated with these energies are:

$$\psi_n = a\,\sin\sqrt{\frac{2mE_n}{\hbar^2}}x = a\,\sin\frac{n\pi}{L}x$$

The constant $a$ is chosen so that the total probability of finding the particle in the region from $x = 0$ to $x = L$ is 1, which gives $a = \sqrt{2/L}$.

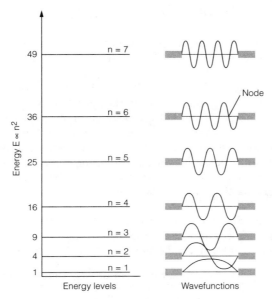

Fig. 2.    *The allowed energy levels and corresponding wavefunctions for a particle in a box.*

The integer $n$ in the equations for the allowed energies and wavefunctions of the particle is an example of a **quantum number**. The permitted energies and the shapes of the associated wavefunctions for the particle in a box are shown in *Fig. 2* for increasing values of $n$. All the wavefunctions except $\psi_1$ possess **nodes** where there is zero probability of finding the particle at these points. Whilst the most probable location for a particle with one quantum of energy is exactly midway between the walls, a particle with two quanta of energy has zero probability of being located at that point.

The separation of adjacent energy levels is:

$$\Delta E = E_{n+1} - E_n = (2n+1)\frac{h^2}{8mL^2}$$

This expression illustrates two general features of quantum mechanical descriptions of physical systems:

(i)    energy levels are more closely spaced as the physical dimension ($L$) of the system increases (*Fig. 3*);

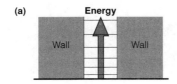

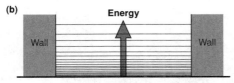

Fig. 3.    *Energy levels are more widely space in (a) a narrow box than (b) a wide box.*

(ii) energy levels are more closely spaced as the mass of the particle ($m$) increases.

For everyday objects the allowed energy levels are so close together that the system can be treated as effectively non-quantized.

**Zero point energy**

The lowest energy a system is allowed to possess corresponds to the lowest quantum number of the system. For the **particle in a box**, this is $n = 1$. Therefore the lowest energy the particle can possess is not zero, as is the case for classical mechanics, but $E_1 = h^2/8mL^2$. This intrinsic, irremovable energy is called the **zero point energy**. The existence of a zero point energy is a direct manifestation of the **uncertainty principle** since if the particle is confined to a finite space its momentum, and hence kinetic energy, cannot be zero. There is no zero point energy for a **particle in a circular orbit**.

**Particle in a circular orbit**

A particle of mass $m$ moving around a circular orbit of radius $r$ with a velocity $v$ has linear momentum $p = mv$ and angular momentum $= mvr$. If the potential energy is zero, then the total energy is entirely kinetic and given by:

$$E = \frac{p^2}{2m} = \frac{\left(\text{angular momentum}\right)^2}{2mr^2} = \frac{\left(\text{angular momentum}\right)^2}{2I}$$

where $I = mr^2$ is called the **moment of inertia** of the particle about the center of its path. The moment of inertia is the rotational equivalent of mass in linear motion. The **de Broglie equation** is used to express the angular momentum of the particle in terms of the wavelength of its associated **wavefunction**:

$$\text{angular momentum} = pr = \frac{h}{\lambda}r$$

hence

$$E = \frac{\left(\frac{h}{\lambda}r\right)^2}{2I}$$

The **boundary condition** of the system is that the shape of the wavefunction must repeat after each circuit of 360° around the circumference of the trajectory along which the particle travels (*Fig. 4*). If this condition is not satisfied then the wavefunction cancels out, or destructively interferes, on each circuit. Therefore physically acceptable wavefunctions must have wavelengths:

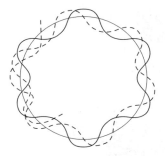

*Fig. 4.*   *An allowed wavefunction for a particle in a circular orbit must repeat after 360° (solid line).*

$$\lambda = \frac{2\pi r}{n} \qquad n = 0,1,2\ldots$$

(The value $n = 0$, corresponding to a wavefunction of constant amplitude, is allowable for a particle on a ring, in contrast to the boundary conditions for the **particle in a box** which require nodes in the wavefunctions at the walls of the box.) The allowed energy quanta for the particle on the ring are therefore:

$$E_n = \frac{\left(\frac{hn}{2\pi}\right)^2}{2I} = \frac{n^2\hbar^2}{2I} \qquad n = 0,\pm1,\pm2\ldots$$

Both positive and negative values of the quantum number are permitted, corresponding to circular motion with the same kinetic energy in either a clockwise or anticlockwise direction. The corresponding allowed quantized values for the angular momentum are:

$$\text{angular momentum} = \frac{h}{\left(\frac{2\pi r}{n}\right)} r = n\hbar \qquad n = 0,\pm1,\pm2\ldots$$

The existence of an $n = 0$ quantum number means that a rotating particle has no irremovable **zero point energy**. This conclusion is consistent with the uncertainty principle. Although the particle is confined to a circle, nothing is known about the particle's position within the whole range of possible angular positions from 0 to 360° so zero angular momentum is possible.

**Degeneracy**

The existence of different states of motion with the same energy is known as **degeneracy**. For the rotating particle all states with $|n| > 0$ are doubly degenerate. The state with $n = 0$ is non-degenerate because in this state the particle is stationary and there is no possibility of different directions of travel.

**Quantum tunneling**

When a particle of energy $E$ is confined by a non-infinite potential barrier $V$, quantum mechanics shows there is still some probability of finding the particle in the region of space on the other side of the barrier, even when $V > E$. In the classical mechanics description the particle has insufficient energy to surmount the barrier and zero probability of existence on the other side. The probability of this **quantum tunneling** decreases as both the height and width of the potential barrier increase (*Fig. 5*).

Tunneling arises because the **wavefunction** does not fall abruptly to zero at

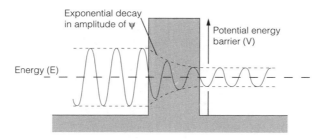

*Fig. 5.   The wavefunction representation of quantum tunneling of a particle through a non-infinite potential barrier.*

the wall (unless the potential is infinite). The maximum amplitude of the wavefunction falls exponentially within the region of space of the potential barrier and a non-zero wavefunction beyond the barrier corresponds to a non-zero chance of finding the particle there. Tunneling probability decreases rapidly with increasing mass of particle, and with the width of the barrier, so it is generally only an important phenomenon for electrons and protons over atomic distances.

# G5 THE STRUCTURE OF THE HYDROGEN ATOM

---

## Key Notes

**Hydrogen spectrum: Rydberg series**

The emission spectrum of a hydrogen atom consists of discrete frequencies, $v$, of light forming the Rydberg series of groups of regular pattern obeying the relationship, $v = R_H (1/n^2_1 - 1/n^2_2)$ with integer values of $n_1$ and $n_2$.

**Interpretation of the hydrogen spectrum**

The solution to the Schrödinger equation for a single electron moving in the attractive Coulombic potential of a positively charged nucleus produces quantized energy levels whose energy values are inversely proportional to the square of an integer quantum number, $n$. The energy difference between pairs of these energy levels exactly accounts for the Rydberg series of the hydrogen atom emission spectrum.

**Atomic quantum numbers and orbitals**

The wavefunction solutions for an electron in an atom are called atomic orbitals. The boundary conditions impose three quantum numbers on the orbitals: principal quantum number, $n$ (1,2,...); orbital angular momentum quantum number, $l$ (0,1... $n - 1$); magnetic quantum number, $m_l$ (–$l$, .. 0,... $l$). All orbitals with the same value of $n$ constitute a shell. Orbitals with different values of $l$ constitute sub-shells of the shell. Orbitals with $l = 0, 1, 2, 3$ are called $s$, $p$, $d$, $f$, orbitals respectively. All orbitals in a sub-shell of a hydrogenic atom are degenerate.

**Shapes of atomic orbitals**

All $s$ orbitals are spherically symmetric about the center of the atom whereas the shapes of $p$ and $d$ orbitals vary with angular direction. The three $p$ orbitals have lobes pointing along the $x$, $y$, and $z$ axes, respectively. The five $d$ orbitals have more complex angular shapes. The radius of maximum probability of electron location in a shell of $s$, $p$, or $d$ orbitals increases with principal quantum number.

**Related topics**

Quantization of energy and
    particle-wave duality (G3)
The wave nature of matter (G4)
Many-electron atoms (G6)

Valence bond theory (H2)
Molecular orbital theory of diatomic
    molecule I (H3)

---

**Hydrogen spectrum: Rydberg series**

A hydrogen atom **emission spectrum** is obtained by passing an electric discharge through a tube of low-pressure hydrogen gas (to form excited hydrogen atoms) and dispersing the emitted light into its constituent wavelengths using a prism or diffraction grating. The resulting spectrum consists of light emitted at discrete frequencies only. The emitted frequencies, $v$, occur in distinct groups with a regular pattern in different regions of the **electromagnetic spectrum** (*Fig. 1*). The frequencies conform to a very simple expression:

$$v = R_H \left( \frac{1}{n_1^2} - \frac{1}{n_2^2} \right) \qquad n_1 = 1, 2, \ldots \qquad n_2 = (n_1 + 1), (n_1 + 2), \ldots$$

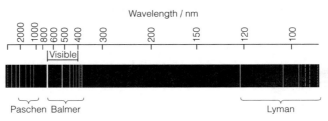

*Fig. 1.   The Rydberg series of the hydrogen atom emission spectrum.*

The sequence is known as the **Rydberg series** and the quantity $R_H = 109\,677$ cm$^{-1}$ is called the **Rydberg constant**. The emission with frequencies corresponding to $n_1 = 1$ is called the **Lyman series** and occurs in the ultraviolet. The **Balmer series** ($n_1 = 2$) occurs in the visible region. The **Paschen, Brackett** and **Pfund series** ($n_1 = 3,4,5$, respectively) are in the infrared.

The existence of discrete spectroscopic frequencies is evidence that the energy of the electron in the hydrogen atom is **quantized**. A photon of light is emitted when the electron moves from a higher to a lower energy level separated by energy difference, $\Delta E = h\nu$.

**Interpretation of the hydrogen spectrum**

The **Rydberg series** of lines in the hydrogen emission spectrum is accounted for by solving the **Schrödinger equation** for the electron in the hydrogen atom. The potential energy $V$ of an electron (one negative unit of elementary charge) at a distance $r$ from a central nucleus of positive charge $Ze$ ($Z$ is the atomic number) is described by the **Coulomb potential**:

$$V = -\frac{Ze^2}{4\pi\varepsilon_0 r}$$

where $\varepsilon_0$ is the vacuum permittivity. For hydrogen, $Z = 1$. The minus sign indicates attraction between the opposite charges of the electron and the nucleus. $V$ is zero when the electron and nucleus are infinitely separated and decreases as the particles approach.

The Schrödinger equation for a single particle moving in this potential energy can be solved exactly. The imposition of appropriate **boundary conditions** (that the **wavefunctions** approach zero at large distance) restricts the system to certain allowed wavefunctions and their associated energy values. The allowed **quantized** energy values are given by the expression:

$$E_n = -\frac{\mu e^4 Z^2}{8\varepsilon_0^2 h^2 n^2} \qquad n = 1,2,\ldots$$

where $\mu = m_e m_n/(m_e + m_n)$ is the **reduced mass** of the electron $m_e$ and nucleus $m_n$. The energy level formula applies to any **one-electron atom** (called **hydrogenic atoms**), e.g. H, He$^+$, Li$^{2+}$, Be$^{2+}$, etc.

The difference between any pair of energy levels in a hydrogenic atom is:

$$\Delta E = E_{n_2} - E_{n_1} = \frac{\mu e^4 Z^2}{8\varepsilon_0^2 h^2}\left(\frac{1}{n_1^2} - \frac{1}{n_2^2}\right)$$

and the values of the physical constants give exact agreement (using appropriate units) with the Rydberg constant derived experimentally from the frequencies of the lines in the hydrogen emission spectrum ($Z = 1$), $\mu e^4/8\varepsilon_0^2 h^2 = hcR_H$.

The distribution of energy levels for the hydrogen atom:

$$E_n = -\frac{hcR_H}{n^2} \qquad n = 1, 2, \ldots$$

is shown in *Fig. 2*. The quantum number $n$ is called the **principal quantum number**. The energies are all negative with respect to the zero of energy at $n = \infty$ which corresponds to the nucleus and electron at infinite separation. The energy of the ground state (the state with the lowest allowed value of the quantum number, $n = 1$) is:

$$E_1 = -hcR_H$$

and is an energy $hcR_H$ more stable than the infinitely separated electron and nucleus. The energy required to promote an electron from the ground state ($n = 1$) to infinite distance from the nucleus ($n = \infty$) is called the **ionization energy, $I$**. For hydrogen, $I = hcR_H = 2.179 \times 10^{-18}$ J, which corresponds to 1312 kJ mol$^{-1}$ or 13.59 eV.

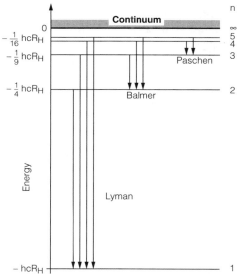

**Fig. 2.** *The energy levels of the hydrogen atom expressed in units of the Rydberg constant $R_H$ relative to a zero energy of infinitely separated proton and electron.*

The energy levels in *Fig. 2* are widely spaced at small values of $n$ but converge rapidly as $n$ increases. At small $n$ the electron is confined close to the nucleus by the electrostatic charge and the energy levels are widely spaced as for a **particle in a narrow box**. At large $n$ the electron has sufficient energy to move at large distances from the nucleus and the energy levels are closer together like those of a **particle in a large box** (see Topic G4).

**Atomic quantum numbers and orbitals**

The full solution of the Schrödinger equation for **hydrogenic atoms** contains three different **quantum numbers** to specify uniquely the different allowed **wavefunctions** and energies of the states of the electron. The **principal quantum number, $n$**, is the only one of the three to appear in the formula for the energy of the various allowed states, and is sufficient to explain the **Rydberg series** of the hydrogen atom emission spectrum.

In addition to the principal quantum number, $n$ (= 1,2,3...) are the **orbital angular momentum quantum number**, $l$, which takes the values, $l = 0,1,2...(n–1)$ and the **magnetic quantum number, $m_l$,** which takes the values, $m_l = –l, –(l – 1),...(l – 1), l$. Therefore, for a given value of $n$ there are $n$ allowed values of $l$, and for a given value of $l$ there are $(2l + 1)$ allowed values of $m_l$. For example, when $n = 2$, $l$ can have the value 0, for which $m_l$ can have the value 0 only, or 1, for which $m_l$ can have the values –1, 0 and 1.

Each wavefunction, which is specified by a unique set of the 3 quantum numbers, is called an **atomic orbital**. All orbitals with the same principal quantum number $n$ are said to belong to the same **shell** of the atom. Orbitals with the same value of $n$ but different values of $l$ are known as the **sub-shells** of the given shell. The sub-shells are usually referred to by the letters $s$ (for sub-shells with $l = 0$), $p$ (for sub-shells with $l = 1$), $d$ (for sub-shells with $l = 2$), $f$ (for sub-shells with $l = 3$) and $g, h, j$...etc. for larger values of $l$, if required. Thus the $n = 2$ shell contains four orbitals (sub-shells), one $s$ orbital and three $p$ orbitals. Electrons that occupy an $s$ orbital are called $s$ electrons. Similarly, electrons can be referred to as $p, d$... electrons.

A fourth quantum number, the **electron spin quantum number, $m_s$,** is required to uniquely specify each electronic wavefunction. This quantum number can take the value $+\frac{1}{2}$ or $-\frac{1}{2}$ (see Topic G6). No two electron wavefunctions can have the same four quantum numbers so each atomic orbital can accommodate a maximum of two electrons. The pattern of allowed combinations of atomic quantum numbers is shown in *Table 1*.

*Table 1. The allowed combinations of quantum numbers for atoms, and the associated nomenclature of the corresponding orbitals. Each orbital can contain a maximum of two electrons, one each with electron spin quantum numbers $+\frac{1}{2}$ and $-\frac{1}{2}$*

| Shell $n$ value | Sub-shell $l$ value | designation | Orbitals $m_l$ values | Maximum no. of electrons per sub-shell | Maximum no. of electrons per shell |
|---|---|---|---|---|---|
| $n = 4$ | $l = 3$ | 4f | –3, –2, –1, 0, 1, 2, 3 | 14 | |
| | $l = 2$ | 4d | –2, –1, 0, 1, 2 | 10 | |
| | $l = 1$ | 4p | –1, 0, 1 | 6 | 32 |
| | $l = 0$ | 4s | 0 | 2 | |
| $n = 3$ | $l = 2$ | 3d | –2, –1, 0, 1, 2 | 10 | |
| | $l = 1$ | 3p | –1, 0, 1 | 6 | 18 |
| | $l = 0$ | 3s | 0 | 2 | |
| $n = 2$ | $l = 1$ | 2p | –1, 0, 1 | 6 | |
| | $l = 0$ | 2s | 0 | 2 | 8 |
| $n = 1$ | $l = 0$ | 1s | 0 | 2 | 2 |

Only hydrogenic atoms have sub-shells that are **degenerate** (of the same energy). For **many-electron atoms** the energy of the orbital (wavefunction) depends on both $n$ and $l$ so each sub-shell has different energy (Topic G6).

**Shapes of atomic orbitals**

In general, the mathematical equation of each atomic **wavefunction** contains a radial part, describing the value of the wavefunction as a function of radial distance from the center of the atom, and an angular part, describing the value of the wavefunction as a function of all angles about the center, i.e., the value of the wavefunction at all points on the surface of the sphere at a given radius, $r$.

For the $1s$ orbital ($n = 1, l = 0, m_l = 0$) the mathematical form of the wavefunction is:

$$\psi \propto e^{-\frac{r}{a_0}}$$

where $a_0$ is a constant known as the Bohr radius. The wavefunction contains no angular dependence so it has the same shape (an exponential decrease) in all directions from the center of the atom. For this reason the 1s orbital is called a **spherically symmetrical** orbital. The shape of the **boundary surface** (within which there is 95% probability of finding the electron) for the 1s orbital is shown in *Fig. 3*.

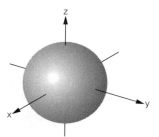

Fig. 3.   *The angular dependence of the boundary surface of the hydrogen 1s orbital.*

For the radial part of the wavefunction, the probability of finding the electron in the region between $r$ and $r + \delta r$ is given by:

**radial probability distribution function** $= 4\pi r^2 \psi^2 \delta r$

where $4\pi r^2 \delta r$ is the volume of the spherical shell of thickness $\delta r$ at radius $r$. A plot of the radial probability distribution function for the 1s orbital is included in *Fig. 4*. The important feature of the radial probability distribution function is that it passes through a maximum. The location of the maximum indicates the most probable radius at which the electron in the orbital will be found. For a hydrogen 1s orbital the maximum occurs at the Bohr radius, $a_0$, which is 53 pm. As with all atomic orbitals, there is zero probability of finding the electron at the nucleus ($r = 0$).

A 2s orbital ($n = 2$, $l = 0$, $m_l = 0$) also has a spherically symmetric wavefunction. However, the radial wavefunction differs from that of the 1s orbital in that

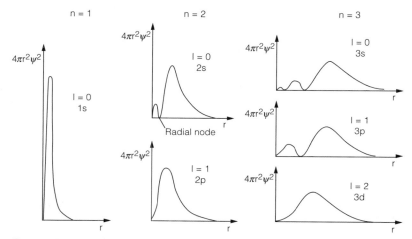

Fig. 4.   *The radial probability density functions for orbitals of the hydrogen atom.*

it passes through zero before it starts to decay to zero at large distances. The corresponding radial probability distribution function (shown in *Fig. 4*) therefore has a radius at which there is zero probability of finding the 2s electron. This is called a **radial node**. The radius at which the 2s electron is most likely to be located is greater than that for the 1s orbital, as expected for an electron possessing greater energy to overcome the nuclear attraction. The pattern repeats for the radial probability distribution of a 3s orbital which has two radial nodes and a yet larger radius for the most probable location of the electron (*Fig. 4*).

The angular wavefunctions of all *p* orbitals (orbitals with $l = 1$) have two lobes pointing in opposite directions with a nodal plane passing through the center of the atom (*Fig. 5*). Consequently, unlike s orbitals, p orbitals are not spherically

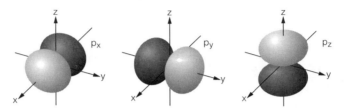

*Fig. 5.    The angular dependence of the boundary surfaces of the hydrogen p orbitals.*

symmetrical and this is an important feature when considering the different types of bonds that can exist between atoms (see Topics H2–H4). A *p* sub-shell consists of three different types of *p* orbital (corresponding to $m_l = -1, 0, +1$) and the three orbitals are normally represented at right angles to each other with the lobes pointing along each of the *x*-, *y*-, and *z*-axes for the $p_x$, $p_y$ and $p_z$ orbitals, respectively. The radial probability distribution function along the axis of each 2*p* orbital does not contain a radial node; the radial probability distribution function of the 3*p* orbital contains one radial node, and so on (*Fig. 4*).

The five *d* orbitals (orbitals with $l = 2$, $m_l = -2, -1, 0, 1, 2$) also have non-spherically symmetric shapes. The boundary surfaces are shown in *Fig. 6*. The *n* = 3 shell is the first shell that contains *d* sub-shells. There is no radial node in the radial probability distribution function along the axis of these orbitals (*Fig. 4*).

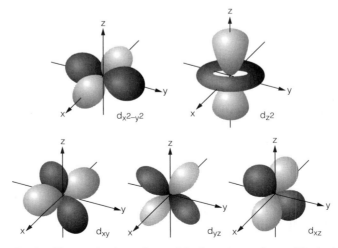

*Fig. 6.    The angular dependence of the boundary surfaces of the hydrogen d orbitals.*

# G6 MANY-ELECTRON ATOMS

## Key Notes

**Electron spin**

An electron possesses ½ unit intrinsic spin angular momentum. The two spin states, representing clockwise or anticlockwise angular momentum, have quantum numbers, $m_s = +½$ and $-½$. Two electrons are termed paired if they have opposite spin.

**Orbital approximation**

In the orbital approximation, the wavefunction of a many-electron atom is approximated as the product of the wavefunctions of individual hydrogenic atomic orbitals, $\psi = \psi(1)\psi(2)\ldots$.

**Penetration and shielding**

Electrons in orbitals between the nucleus and the electron under consideration repel the electron under consideration and reduce the net attraction it experiences from the nucleus. The effect is called shielding and is accounted for by reducing the actual nuclear charge $Ze$ to an effective nuclear charge $Z_{eff}e$. Electrons in $s$, $p$, $d$, $f$ orbitals are successively better shielded by core electrons (and consequently have higher energy) because the radial probability functions of these orbitals successively penetrate less towards the center of the atom.

**Pauli exclusion principle**

No more than two electrons can occupy a given atomic (or molecular) orbital, and when two electrons do occupy one orbital their spins must be paired.

**Electron configuration**

The electron configuration is the distribution of electrons into individual atomic orbitals. The electron distribution of lowest energy is called the ground state configuration.

**Aufbau principle**

The Aufbau (building-up) principle is the process of filling atomic orbitals (sub-shells) with electrons in the order of increasing sub-shell energy. Each individual atomic orbital (with a unique set of $n$, $l$, $m_l$ quantum numbers) can accommodate a maximum of two electrons with paired spins.

**Single occupancy rule**

Electrons singly occupy degenerate orbitals of a sub-shell before pairing into the same orbital, in the ground state configuration.

**Hund's rule**

The ground electron state configuration contains the maximum number of unpaired electrons.

**Atomic term symbols**

The atomic term symbol is a short-hand notation that represents an electron configuration of an atom. For a total spin angular momentum quantum number, $S$, and total angular momentum quantum number, $L$, the term symbol is wirtten as $^{2S+1}\{L\}$, where $\{L\}$ is the letter $S$, $P$, $D$, $F\ldots$ for $L = 0, 1, 2, 3\ldots$, respectively. The quantity $2S+1$ is called the multiplicity.

**Related topics**

The structure of the hydrogen atom (G5)

Chemical and structural effects of quantization (G7)

Valence bond theory (H2)

Molecular orbital theory of diatomic molecules I (H3)

Molecular orbital theory of diatomic molecules II (H4)

**Electron spin**

Every electron possesses an intrinsic ½ unit of angular momentum. This is a fundamental property of the electron, like its mass and charge, that cannot be altered. The spin angular momentum may be clockwise or anticlockwise corresponding to two quantum states with **electron spin quantum number**, $m_s = +\frac{1}{2}$ and $-\frac{1}{2}$. The two spin states of an electron are often represented by the symbols ↑ and ↓, respectively. When electron spins are **paired** (↑↓) there is zero net spin angular momentum because the spin angular momentum of one electron is cancelled by the opposite spin angular momentum of the other electron.

**Orbital approximation**

In principle, wavefunctions for a **many-electron atom** (atoms containing two or more electrons) describe the behavior of all the electrons simultaneously. However, the **Schrödinger equation** for such atoms cannot be solved exactly because each electron interacts with every other electron as well as with the nucleus. In the **orbital approximation**, the many-electron wavefunction is described as the product of the wavefunctions of the individual atomic orbitals occupied by each electron in the atom, $\psi = \psi(1)\psi(2)\ldots$

Each individual orbital can be considered like a **hydrogenic atomic orbital** with the potential energy modified by the effect of the other electrons in the atom.

**Penetration and shielding**

An electron in a many-electron atom experiences Coulombic repulsion from all the other electrons present. The extent of repulsion can be represented as a negative charge at the nucleus which cancels out a proportion of the $Z$ units of positive charge from the protons in the nucleus ($Z$ is the **atomic number** of the atom). The cancelling out reduces the charge of the nucleus from $Ze$ to $Z_{eff}e$, called the **effective nuclear charge**. The other electrons are described as **shielding** the nuclear charge. The effective nuclear charge is a convenient way of expressing the net effect of the attraction of the electron to the nucleus and its repulsion from all other electrons by means of a single equivalent charge at the center of the atom. For example, $Z_{eff}$ experienced by an electron in the 2s orbital of Li ($Z = 3$) is 1.26 indicating that the two core electrons do not provide complete screening of two units of positive charge.

The effective nuclear charges experienced by electrons in $s$, $p$, $d\ldots$ orbitals are different because the shapes of their wavefunctions are different. The **radial probability distribution function** (Topic G5) for an $s$ orbital shows there is greater probability of finding the electron at distances close to the nucleus than for an electron in a $p$ (or $d\ldots$) orbital of the same shell (*Fig. 1*). The $s$ electron has

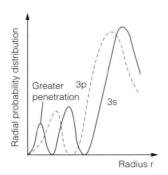

*Fig. 1.    Comparison of penetration close to the nucleus for an s and p orbital.*

greater **penetration** through the inner shells than the $p$ (or $d$...) electron. Consequently, an $s$ electron experiences less shielding from electrons in inner shells, a greater $Z_{eff}$, and is more tightly bound (has lower energy). Therefore, in general, the energies of orbitals in the same shell of a many-electron atom increase in the order, $s < p < d < f$. This explains why, in contrast to **hydrogenic** atoms, the **sub-shell** orbitals of many-electron atoms are not **degenerate**. However, the individual orbitals of a particular sub-shell, as specified by the magnetic quantum number $m_l$ (for example the three $p$ orbitals), remain degenerate because each one has the same radial probability function and therefore experience the same effective nuclear charge.

**Pauli exclusion principle**

The **Pauli exclusion principle** is a fundamental outcome of quantum mechanics. It states that

*no more than two electrons can occupy a given atomic orbital (or molecular orbital), and when two electrons do occupy one orbital their spins must be paired.*

The exclusion principle is a key feature in the derivation of **electron configuration** through the **Aufbau** (or **building-up**) **principle**.

**Electron configuration**

The distribution of electrons into individual atomic orbitals within the **orbital approximation** is called the **electron configuration** of the atom. The electron distribution of lowest energy is called the **ground state** configuration. All electron distributions of higher energy are called **excited state** configurations. For example, the ground state of a hydrogen atom consists of a single electron in a $1s$ orbital, so its configuration is written $1s^1$. The configuration of the ground state of the helium atom is written $1s^2$. The first excited state of helium requires energy to promote one of the $1s$ electrons into the $2s$ orbital and the configuration is $1s^1 2s^1$.

**Aufbau principle**

The procedure used to determine the **ground state configurations** of many-electron atoms by sequentially filling up individual atomic orbitals is called the **Aufbau** (or **building-up**) principle. Building-up starts with the lowest energy orbital first and continues until all electrons in the atom (equal to the **atomic number**, $Z$) have been assigned to an orbital. The order of occupation of sub-shells is $1s$ $2s$ $2p$ $3s$ $3p$ $4s$ $3d$ $4p$ $5s$ $4d$ $5p$ $6s$ $5d$ $4f$ $6p$ ....

The order of occupation is approximately the order of energy of the individual orbitals as determined by **principal shell** and **penetration**. The ability of the $4s$ orbital to penetrate the inner shells lowers its energy below that of the $3d$ orbitals. The same occurs for the $5s$ orbital.

Each orbital within a sub-shell may accommodate a maximum of two electrons (**Pauli exclusion principle**). An $s$ sub-shell contains one $s$ orbital and is complete when it contains two electrons. A $p$ sub-shell contains three degenerate $p$ orbitals and is complete when it contains six electrons. A $d$ sub-shell contains five degenerate $d$ orbitals and is complete when it contains ten electrons (see Topic G5, *Table 1*).

As an example of the Aufbau principle, consider the six electrons of the carbon atom ($Z = 6$). Two electrons fill the $1s$ orbital, two electrons fill the $2s$ orbital and the last two electrons occupy the $2p$ orbitals. The configuration is therefore $1s^2 2s^2 2p^2$. Since $1s^2$ represents a filled **principal quantum shell**, equivalent to a He atom, the configuration can also be written as $[He]2s^2 2p^2$. A filled principal shell is called a **closed shell**. Electrons which occupy orbitals in the outermost principal shell (called the **valence shell**) are called **valence electrons**; electrons which occupy closed inner shells are called **core electrons**. Carbon has four valence electrons occupying $2s$ and $2p$ orbitals. Sodium ($Z = 11$) has an electron configuration $1s^2 2s^2 2p^6 3s^1$ (also written as [Ne]$3s^1$) and has one valence electron. All alkali metals have similar configurations of a single $s$ electron outside a filled core and such repeating patterns in electronic structure form the basis for **periodicity** in the elements (see Topic G7).

**Single occupancy rule**

The two $2p$ electrons in the [He]$2s^2 2p^2$ ground state electron configuration of carbon must be distributed amongst three **degenerate** $2p$ orbitals. Each electron occupies a separate $p$ orbital in order to minimize electrostatic repulsion between the electrons. The $p$ orbitals point in different directions in space (see Topic G5) so electrons in separate orbitals are, on average, further apart from each other than if they both occupied the same orbital. The ground state electron configuration of carbon ($Z = 6$) is therefore more precisely written as [He]$2s^2 2p_x^1 2p_y^1$ and that for nitrogen ($Z = 7$) as [He]$2s^2 2p_x^1 2p_y^1 2p_z^1$, and so on. It is a general rule of the Aufbau principle that electrons occupy different orbitals of a given sub-shell singly before double occupancy.

**Hund's rule**

Hund's rule states that

*the ground state electron configuration of an atom maximizes the number of unpaired electrons.*

For example, the **Augbau principle** shows that the ground state configuration of carbon is [He]$2s^2 2p_x^1 2p_y^1$. Since the two $2p$ electrons occupy different $2p$ orbitals both electrons can have a spin quantum number, $m_s$, of $+\frac{1}{2}$ (or both of $-\frac{1}{2}$) without violating the **Pauli exclusion principle**. Hund's rule dictates that the two $2p$ electrons in ground state C have the same (unpaired) spin. Similarly, all three $2p$ electrons in the ground state of $N$ have the same spin. The ground state electron configurations of the first row elements is shown in *Fig. 2*.

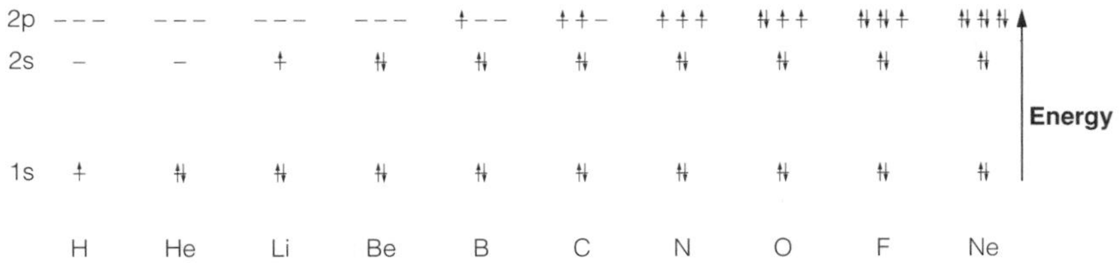

*Fig. 2.    Ground state electron configurations of the first row elements. The representation is schematic only and does not indicate relative energy level separation.*

Hund's rule arises through a quantum mechanical phenomenon known as **spin correlation** in which electrons with the same spin tend to stay further apart from each other, on average, than electrons with opposite spins.

**Atomic term symbols**

The **electron configuration** (ground or excited) of any atom may be succinctly represented by an appropriate **atomic term symbol**. It takes the form:

$$^{2S+1}\{L\}$$

where $S$ is the **total spin angular momentum quantum number** and $\{L\}$ is a letter that signifies the **total orbital angular quantum number**, $L$. Thus, when

$L = 0, 1, 2, 3. . .$ the corresponding letter used in the term symbol is $S, P, D, F, . . .$ Note the comparative symbolism with the $s, p, d, f. . .$ letter used to represent electrons in individual **atomic orbitals** of **orbital momentum quantum number** $l = 0, 1, 2, 3. . .$, respectively (Topic G5).

The quantity $2S+1$ is known as the **multiplicity**. Two electrons with paired spin contribute no net spin angular momentum. Thus for an atom containing no unpaired electrons, $S = 0$, and the multiplicity is 1 (a **singlet state**); for an atom containing one unpaired electron, $S = ½$, and the multiplicity is 2 (a **doublet state**); for an atom containing two unpaired electrons, $S = 1$, and the multiplicity is 3 (a **triplet state**), and so on.

For example the ground state electron configuration of He is $1s^2$ and has term symbol $^1S$. The excited configuration of He in which an electron is promoted from the $1s$ orbital to the $2s$ orbital, and spins are opposite in each orbital, also has term symbol $^1S$. However, if the electron spins are parallel in this electron configuration, the term symbol is $^3S$. Similarly, the $1s2p$ excited state of He gives rise to term symbols $^1P$ and $^3P$, depending on whether the electron spins are paired or unpaired. The term symbol for the ground state of Li $(1s^22s^1)$ is $^2S$.

In more detailed treatments the term symbol also includes a subscript number after the symbol for $\{L\}$ to indicate the value of the total electronic angular momentum (spin and orbital) quantum number $J$.

# G7 CHEMICAL AND STRUCTURAL EFFECTS OF QUANTIZATION

## Key Notes

**Periodicity and the periodic table**

The structure of the periodic table reflects the ground state electronic configuration of the elements according to the Aufbau principle of allocating electrons to atomic orbitals in order of increasing orbital energy. The elements are arranged in periods such that every element in a group has the same configuration of valence electrons in the outer shell. The analogous valence configuration gives rise to the periodicity in physical and chemical properties of the elements.

**Atomic radii**

The atomic radius is a measure of the size of an atom. Atomic radii decrease across a period, because of the increase in nuclear charge and incomplete nuclear shielding, and increase down a group, because successively larger principal shells are occupied.

**Ionization energies**

The first (and second…) ionization energies are the minimum energies required to remove the first (and second…) electrons to infinite distance from the atom. Ionization energies generally increase across a period, because the outer electron becomes more tightly bound as nuclear charge increases, and decrease down a group, because the outer electron occupies successively larger orbitals more weakly bound to the nucleus.

**Atomic transitions and selection rules**

An atomic transition is the movement of electron(s) between two electronic configurations (or states). The difference in energy between the two states determines the frequency of associated radiation. Allowed atomic electronic transitions obey the selection rule, $\Delta l = \pm 1$ and $\Delta m_l = 0, \pm 1$.

**Spectra of hydrogen-like atoms**

The alkali metal atoms contain a single $s$ valence electron outside a closed core and are called hydrogen-like. The energy level distribution in hydrogen-like atoms, and the pattern of the emission spectra, closely resemble those of the hydrogen atom.

**Related topics**

The structure of the hydrogen atom (G5)

Many-electron atoms (G6)

General features of spectroscopy (I1)

**Periodicity and the periodic table**

Elements in the periodic table are arranged in order of increasing **atomic number**, $Z$. The electrons of each element (equal in number to $Z$) fill the **atomic orbitals** in order of increasing energy, according to the rules of the **Aufbau (or building-up) principle** (see Topic G6). The locations in the periodic table which correspond to the sequential filling up of different types of orbital are shown in *Fig. 1*. The structure of the table reflects the recurrence of analogous **electron**

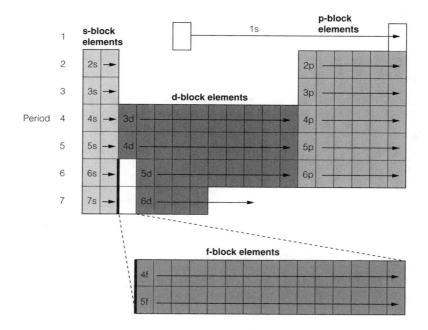

Fig. 1.    The structure of the periodic table in terms of the filling up of atomic orbitals.

configurations; every element in a group (column) has the same configuration of valence electrons.

Each period (row) starts with an element that has one s electron in a new principal quantum shell (the alkali metals) and ends with an element that has completely filled, or closed, sub-shells (the noble gas group). The first period is only two elements long since the 1s orbital can accommodate only two electrons. The second period contains two s-block elements and six p-block elements. The third period ends after the 3p elements, rather than continuing with the 3d elements, because of the effects of shielding and penetration (Topic G6).

The analogous electron configurations confer the periodicity on the physical and chemical properties of the elements on which the structure of the periodic table was originally based.

**Atomic radii**

An atomic radius is defined in terms of the internuclear distance between bonded atoms in solids and molecules. The atomic radii of the elements as a function of atomic number are shown in Fig. 2 and show trends that match the periodicity of the periodic table. In general, atomic radii decrease from left to right across a period and increase down each group. The atomic radii increase down a group because the valence electrons are in successively larger principal shells of higher energy and greater average distance from the center of the atom. The decrease in radii across a period is caused by the increase in nuclear charge which attracts the electron orbitals more closely to the nucleus. Although the increase in nuclear charge is partly cancelled by the increase in the number of electrons, the shielding of the nuclear charge is incomplete.

**Ionization energies**

The first ionization energy is the minimum energy required to remove the most weakly bound electron to infinite distance from a many-electron atom. The

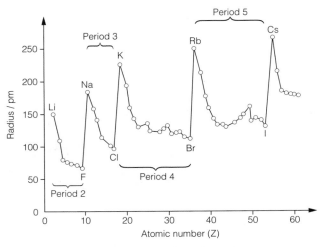

*Fig. 2.   Atomic radii as a function of atomic number.*

second ionization energy is the minimum energy required to remove the next electron from the singly charged cation, and so on.

The variation of first ionization energy with **atomic number**, $Z$, is shown in *Fig. 3*. The general trend is for ionization energy to increase along a period and decrease down a group. The decrease down a group arises because the outer-most electron occupies a successively larger **principal quantum shell** and is therefore less tightly bound to the nucleus. Ionization energy generally increases along a period because the outermost electron belongs to the same principal shell, but nuclear charge increases so the electrons are more tightly bound. The alkali metals (Group I) have lowest first ionization energies because the single $s$

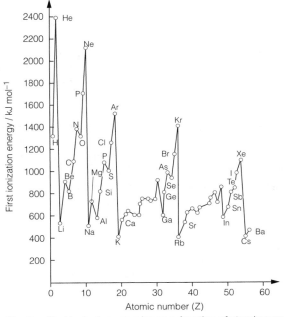

*Fig. 3.   First ionization energies as a function of atomic number.*

electron occupies a new principal shell and the nuclear charge is well-shielded by the complete noble-gas-like core.

The kink in the ionization energy trend between the elements Be and B in the second period, and between Mg and Al in the third period, arises because the outermost electrons in B and Al occupy $p$ orbitals, which are less strongly bound (less **penetration** into the core), than the electrons in the $s$ orbitals of the preceding elements. The additional slight kink in ionization energy between N and O is due to the extra electron–electron repulsion that occurs when one of the $2p$ orbitals becomes doubly occupied. This effect is less pronounced between P and S in period 3 because their orbitals are larger and more diffuse.

**Atomic transitions and selection rules**

Each electron configuration of an atom (also called a **state** or a **level**) possesses a well-defined energy. The movement (or **transition**) of electrons between any pair of states is associated with a quantum of energy exactly equal to the difference in energy between the initial and final state, $\Delta E = h\nu$. This is the basis of **atomic spectroscopy**. Examples include the **Rydberg series** in the hydrogen atom emission spectrum (see Topic G5) or the spectra of **hydrogen-like** atoms (this Topic).

Not all transitions between all possible pairs of energy levels are **allowed**. The **selection rule** for atomic transitions is, $\Delta l = \pm 1$ and $\Delta m_l = 0, \pm 1$. The selection rules arise because angular momentum must be conserved. A photon of electromagnetic radiation has one unit of angular momentum so the angular momentum of the electron involved in the transition must change by one unit whenever a photon is emitted or absorbed. Therefore an $s$ electron ($l = 0$) cannot make a transition into a $d$ orbital ($l = 2$), or *vice versa*, because the photon cannot provide (or carry away) two units of angular momentum.

**Spectra of hydrogen-like atoms**

The electron configurations of all alkali metal atoms consist of a single electron in an $s$ orbital surrounding filled core shells, e.g. Li ($Z = 3$), Na ($Z = 11$), have electron configurations [He]$2s^1$, [Ne]$3s^1$, etc. These atoms are termed **hydrogen-like** since the single valence electron is comparatively unaffected by the core electrons and behaves similarly to a **hydrogenic atom** in which a single electron experiences a Coulomb potential only from a positive core. The shielding of the nuclear charge by the core electrons is incorporated into the energy level formula of hydrogenic atoms using an **effective nuclear charge** $Z_{eff}e$. The energy level diagram for Li is shown in *Fig. 4* and is similar to the energy level diagram for hydrogen (Topic G5) except that the different **penetration** of **sub-shells** into the core removes the **degeneracy** of the sub-shells.

The similar pattern of energy levels of alkali metal atoms produces spectra similar in appearance to the **Rydberg series** of the hydrogen atom emission spectrum (Topic G5). Transitions allowed by the atomic **selection rule** are marked on *Fig. 4*. The series of frequencies corresponding to transitions between different sub-shells were historically called **sharp**, **principal**, **diffuse** and **fine**. The first letters of these descriptive terms account for the modern nomenclature of the different types of **atomic orbital**: $s$, $p$, $d$ and $f$. The energy levels giving rise to each series are labeled S, P, D and F. The transition from the lowest level of the P series to the lowest level of the S series for the alkali metals gives emission in the visible region of the spectrum. This light is responsible for the characteristic colored flame test for these atoms.

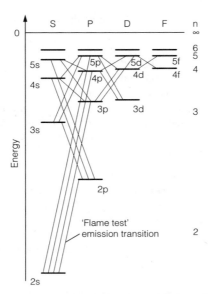

Fig. 4.    Energy level diagram for lithium showing some allowed electronic transitions.

# G8 STATISTICAL THERMODYNAMICS

---

## Key Notes

**Statistical thermodynamics**

Statistical thermodynamics attempts to both qualitatively and quantitatively explain measurable properties obtained through classical thermodynamics (entropy, heat capacity, etc.) by analyzing the quantized behavior of systems at the molecular level.

**The Boltzmann distribution law**

The Boltzmann distribution is a statistical approach which describes the distribution of the components of a system, molecules or atoms, for example, over the available states of that system. The most probable configuration of any system is given when the population of each state, $n_i$, is given by the Boltzmann law:

$$\frac{n_i}{N} = \frac{e^{-\varepsilon_i/k_BT}}{\sum_{i=0}^{\infty} e^{-\varepsilon_i/k_BT}}$$

By defining the lowest energy level to be equal to zero, the population in a level of energy $\varepsilon_j$ above this is given by $n_j = n_o g_j \exp(-\varepsilon_j/k_BT)$ where $g_j$ is the degeneracy of level j.

**The partition function**

The partition function, $q$, is defined as:

$$q = \sum_j g_j e^{-\varepsilon_j/k_BT}$$

$q$ is a temperature-dependent dimensionless number which provides a measure of the ability of molecules to access energy levels above the ground state. The larger the value of $q$, the more molecules access energy levels above $\varepsilon_o$. $q$ varies from 1 at absolute zero ($n_o = N$) to an exceedingly large value where the energy levels are closely spaced and at high temperature ($n_o \to 0$).

**Separation of the molecular partition function**

The total partition function for a molecule is obtained from the product of the individual terms: $q = q_{trans} \cdot q_{rot} \cdot q_{vib} \cdot q_{elec}$. Each partition function may be calculated from knowledge of the energy spacings of the individual terms. $q_{elec}$ is approximately equal to 1 for most materials, with the remaining terms being given by the relationships: $q_{trans} = (2\pi m k_B T/h^2)^{1/2}V$, $q_{rot}$ (diatomic molecule) $= 8\pi^2 I k_B T/\sigma h^2$ and $q_{vib} = (1-e^{-h\nu/k_BT})^{-1}$.

**Thermodynamic parameters and the partition function**

The partition function may be used for direct calculation of the values of thermodynamic parameters, the most significant of which is the entropy: $S = k_B \ln q^N + U/T$. The specific case of a perfect monatomic gas yields the Sackur-Tetrode equation: $S = nR \ln[e^{5/2}(2\pi m k_B T/h^2)^{3/2}(k_B T/p)]$.

**Heat capacity**

Because the thermal component of the energy of a system may be calculated from the partition function, the heat capacity may be calculated from the differential of this value with respect to time. The maximum molar heat capacity is equal to $R/2$ for each degree of freedom, that is, each independent mode of motion. Thus, a gaseous diatomic molecule may have three translational degrees of freedom (one for each orthogonal direction of motion), two rotational degrees of freedom (from rotation about each of two equatorial axes). One vibrational degree of freedom contributes $R$ to the molar heat capacity – $R/2$ from each of the potential and kinetic energy components.

**Related topics**     The first law (B1)                    Entropy and change (B5)
                       Entropy (B4)                         Free energy (B6)

---

**Statistical thermodynamics**

Classical thermodynamics neither requires, nor takes account of, the molecular nature of matter, whereas chemists are interested in the molecular nature of matter and its properties. Statistical thermodynamics has been a highly successful approach to bridging the gap between the quantized, molecular properties of a system and its macroscopic thermodynamic properties. It is a fundamental premise of statistical thermodynamics that the microscopic properties of a system directly influence those properties which are observable and measurable at the macroscopic level (heat capacity or entropy, for example). Statistical thermodynamics operates effectively because the microscopic properties of a system can be described by focusing only on the most probable molecular state.

Furthermore, since nature places no weighting on any particular one of a set of states of equal energy, the most probable states are those which can be generated in the greatest number of ways. Once the statistical properties of the most probable state have been ascertained, it is then possible to use this information to describe the macroscopic thermodynamic properties of the system in terms of experimentally measurable quantities.

**The Boltzmann distribution law**

The **Boltzmann distribution** is a statistical description of the manner in which the molecules in a system are distributed over the available states of that system.

The standard approach is to consider $N$ molecules in a system of total energy, $E$. Within the system, each molecule may be in one of a number of states, each with energies $\varepsilon_0$, $\varepsilon_1$, $\varepsilon_2$, $\varepsilon_3$, etc. $\varepsilon_0$ is the ground state and the subsequent terms have increasingly higher energies. Each energy level, $\varepsilon_i$, is occupied by $n_i$ molecules. The **total number criterion** recognizes that the total number of molecules is the sum of the number of molecules in each state:

$$\sum_i n_i = N$$

The total energy of molecules in state $\varepsilon_i$ is equal to $\varepsilon_i n_i$, and the **total energy criterion** follows from this:

$$\sum_i \varepsilon_i n_i = E$$

The huge number of possible ways in which the molecules can be distributed across the available energy levels is known as the **configuration** of the system, and must comply with the total energy and total number criteria. For each configuration there are a number of ways, $W$, in which the molecules can be distributed amongst the available energy levels, given by:

$$W = \frac{N!}{n_1! n_2! n_3! n_4! \ldots}$$

The macroscopic properties of the system will be the result of the total configuration of the molecules in the system. The most probable configuration of the system will be that with the largest value of $W$. The state with the maximum value of $W$, which complies with the total energy and total number criteria, is obtained by a straightforward but lengthy calculation. It is found that the maximum $W$ is obtained when the population of each state, $n_i$, is given by the **Boltzmann law**:

$$\frac{n_i}{N} = \frac{e^{-\varepsilon_i/k_B T}}{\sum\limits_{i=0}^{\infty} e^{-\varepsilon_i/k_B T}}$$

If each energy level, $\varepsilon_j$, has a **degeneracy** of $g_j$, then the Boltzmann law may be rewritten in terms of energy levels, rather than states:

$$\frac{n_j}{N} = \frac{g_j e^{-\varepsilon_j/k_B T}}{\sum\limits_{j} g_j e^{-\varepsilon_j/k_B T}}$$

The population ratio between two energy levels, $\varepsilon_i$ and $\varepsilon_j$ is then given by:

$$\frac{n_i}{n_j} = \frac{g_i e^{-\varepsilon_i/k_B T}}{g_j e^{-\varepsilon_j/k_B T}} = \frac{g_i}{g_j} e^{-(\varepsilon_i - \varepsilon_j)/k_B T}$$

If the lowest energy level has a degeneracy of one, and is regarded as having an energy of zero, then it follows that the population $n_j$ of an energy level, $\varepsilon_j$ relative to that of the lowest level $n_o$ is given by:

$$n_j = n_o g_j \exp(-\varepsilon_j/k_B T)$$

at low temperature, $\varepsilon_j \gg k_B T$, and the number of molecules with energies above that of the ground state approaches zero.

**The partition function**

The denominator in the Boltzmann law is of considerable importance in statistical thermodynamics, and is referred to as the **partition function, $q$**.

$$q = \sum_j g_j e^{-\varepsilon_j/k_B T}$$

The significance of $q$ is seen when the total number of molecules is summed over all energy levels:

$$N = \sum_j n_j = \Sigma n_o g_j \exp\left(-\varepsilon_j/k_B T\right) = n_o \Sigma g_j \exp\left(-\varepsilon_j/k_B T\right) = n_o q$$

Hence, $q = N/n_o$.

$q$ is a temperature-dependent dimensionless number. Because $q$ is the reciprocal of the fraction of molecules in the ground state, it provides a measure of the ability of molecules to access energy levels above the ground state. The larger the value of $q$, the more molecules access energy levels above $\varepsilon_o$ (*Fig. 1*). The value of $q$ varies from 1 at absolute zero ($n_o = N$) to an exceedingly large value where the energy levels are closely spaced and at high temperature ($n_o \to 0$).

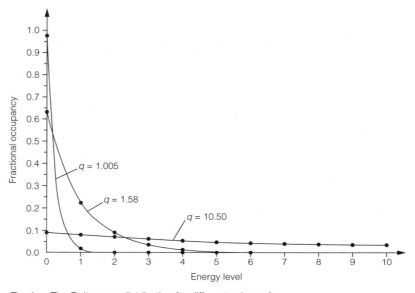

Fig. 1.    *The Boltzmann distribution for different values of q.*

**Separation of the molecular partition function**

The energy levels in a molecule may be separated into several components. Generally:

$$E = E_{trans} + E_{rot} + E_{vib} + E_{elec}$$

representing the sum of the translational, rotational, vibrational and electronic energy components energies respectively. A partition function is associated with each energy term, but as these represent probabilities, the corresponding total partition function is given by the product of the individual terms:

$$q = q_{trans} \cdot q_{rot} \cdot q_{vib} \cdot q_{elec} \qquad or \qquad \ln(q) = \ln(q_{trans}) + \ln(q_{rot}) + \ln(q_{vib}) + \ln(q_{elec})$$

The individual partition functions are calculated from knowledge of the energy spacings of the individual terms. For a diatomic molecule of mass, $m$, with a moment of inertia, I, and vibrational energy spacing $h\nu$ in a container of volume V, the values of the partition functions are given in *Table 1*.

Because the electronic energy levels are widely separated in most materials the **electronic partition function** can be explicitly calculated. However, with few exceptions, electronically excited states in molecules are thermally inaccessible and $q_{elec} \approx 1$.

**Thermodynamic parameters and the partition function**

It is possible to directly relate the partition function to the thermodynamic parameters of a system. The thermally sourced internal energy, $U$, and the entropy, $S$, are given by:

$$U = Nk_B T^2 \partial(\ln q)/\partial T \qquad and \qquad S = k_B \ln q^N + U/T$$

Table 1.   Partition functions for a diatomic molecule

| Property | Partition function | Notes |
|---|---|---|
| Translational partition function | $q_{trans} = (2\pi m k_B T/h^2)^{1/2} V$ | Based on the energy levels for a particle in a box (Topic G4) |
| Rotational partition function | $q_{rot} = 8\pi^2 I k_B T/\sigma h^2$ | Assumes a rigid rotor. $\sigma = 1$ for heteronuclear molecules, such as HF or HCl, and $\sigma = 2$ for homonuclear molecules such as $H_2$, $I_2$, etc. |
| Vibrational partition function | $q_{vib} = (1 - e^{-h\nu/k_B T})^{-1}$ | Assumes a harmonic oscillator in which only the lowest energy vibrational modes are thermally accessible |

For the special case of a monatomic gas, the only contribution to the partition function results from translational energy levels. This ultimately yields the **Sackur-Tetrode equation** for the entropy of a perfect monatomic gas of mass, $m$, at a pressure, $p$:

$$S = nR \ln\left[ e^{5/2} \left(2\pi m k_B T/h^2\right)^{3/2} \left(k_B T/p\right)\right]$$

**Heat capacity**

Partition functions allow calculation of the heat capacity of a system. The following discussion of heat capacity applies to the **constant volume heat capacity**, from which the constant pressure heat capacity may be easily calculated (Topic B1). For a gas, substitution of $q_{trans}$ into the expression for $U$ yields

$$E_{trans} = 3RT/2$$

Therefore the molar translational heat capacity is given by $C_{trans} = dE_{trans} / dT = 3R/2$, and $q_{rot}$ and $q_{vib}$ can be likewise treated. It is found that, for a diatomic gas, both the molar quantities $C_{rot}$ and $C_{vib}$ vary between 0 and R depending upon the ratio of $kT$ to the difference between energy levels, $h\nu$. For $C_{rot}$ or $C_{vib}$, when $h\nu \gg kT$, the heat capacity is zero, rising to a molar value of R when $h\nu \ll kT$. Generally, for a translation or a rotation, the maximum heat capacity is equal to $R/2$ for each **degree of freedom**, that is, each independent mode of motion. Thus, a gaseous diatomic molecule may have three translational degrees of freedom (one for each orthogonal direction of motion), two rotational degrees of freedom (from rotation about each of two axes perpendicular to the main axis of the molecule and to one another). Any vibration contributes $R$ to the

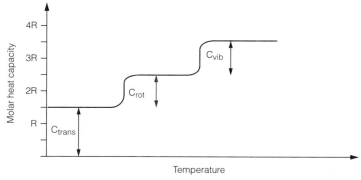

Fig. 2.   Idealized molar heat capacity behavior as a function of temperature for a diatomic molecule.

molar heat capacity, $R/2$ each from the kinetic and potential energy components of the vibration.

The difference between energy levels follows the trend $h\nu_{trans} \ll h\nu_{rot} \ll h\nu_{vib}$, and this means that at low temperature only the translational motion makes a significant contribution to the heat capacity. As temperature increases, the heat capacity progressively increases also, as first the rotational modes, and then the vibrational modes contribute to the heat capacity (*Fig. 2*).

# H1 ELEMENTARY VALENCE THEORY

## Key Notes

**Valence theories**

Valence theories attempt to describe the number, nature, strength and geometric arrangements of chemical bonds between atoms. Although they have been superseded by more sophisticated theories, Lewis theory and VSEPR theory provide two complementary approaches to bonding which remain useful for elementary descriptions of simply bonded molecules.

**Lewis theory**

Lewis theory is a primitive form of valence bond theory, with atoms forming bonds by sharing electrons. No attempt is made to describe the three-dimensional geometric shape of the molecule. The main group elements tend to adopt inert gas electron configurations (octets), although some elements, such as boron or beryllium are energetically stable with incomplete octets. Many larger elements display hypervalency, where it is energetically favorable for more than eight valence electrons to be held in an expanded octet.

**VSEPR theory**

Valence shell electron pair repulsion (VSEPR) theory explains the shapes of molecules by focusing on the bonding orbitals around each atom in isolation. VSEPR dictates that the geometry which maximizes the distances between the electron pairs in the orbitals is adopted. The basic geometry from the minimization of electron–electron repulsion is modified by the differing repulsion strengths of bonding and non-bonding pairs. In ammonia, $NH_3$, for example, there are four valence shell pairs, giving an underlying tetrahedral geometry, but the greater repulsive effect of the non-bonding pair forces the bonding pairs closer to one another than in the ideal tetrahedral geometry.

**Related topics**

Many electron atoms (G6)
Valence bond theory (H2)

Molecular orbital theory of diatomic molecules I (H3)

**Valence theories**

**Valence theories** attempt to describe the number, nature and strength of chemical bonds between atoms. It also describes the geometric arrangement of the bonds, and so the shapes of molecules. The more sophisticated valence theories yield information about the electrical, magnetic, and spectroscopic properties of molecules.

Elementary valence theories invoke two principal bond types. In **ionic bonding**, electrostatic interactions generate bonds between ions formed by electron transfer from one element to the other. In **covalent bonding** two elements are held together by shared electrons in order that both may adopt an energetically favorable electron configuration. In reality, both are extreme forms of the same bonding phenomenon. Pure covalent bonds are formed by elements with identical electronegativities, with more ionic bonding character being introduced to the bond as the electronegativity difference between the elements increases (see Topic H4). Even in extreme cases of ionic bonding, the degree of covalent character may still be quite high.

Two complementary theories were originally developed to explain the number and nature of covalent bonds (Lewis theory) and the shapes of molecules (VSEPR theory). More sophisticated theories have superseded these approaches for detailed investigations, but they remain useful in semi-empirical and non-rigorous discussions of molecular bonding.

**Lewis theory**

The Lewis theory of covalent bonding may be regarded as an elementary form of valence bond theory. It is nonetheless useful for describing covalent molecules with simple covalent bonds, and works successfully in describing the majority of, for example, organic compounds. Lewis theory recognizes both the free energy gains made in the formation of complete atomic electron shells, and the ability of atoms to achieve this state by sharing electrons. The sharing process is used as a description of covalent bonds.

The atoms are firstly drawn so as to represent their relative arrangement, with electron pairs (marked as pairs of dots) between neighboring atoms to indicate a shared bonding electron pair. No attempt is made to describe the three-dimensional geometric shape of the molecule. Multiple bonds are represented by two or three electron pairs as appropriate (*Fig. 1a*). Further electrons are added to each atom, so as to represent the non-bonding electrons and so

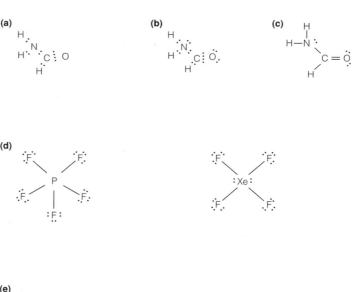

*Fig. 1.    (a)–(c) Development of a Lewis bonding scheme for HCONH₂. Examples of (d) hyper-valency, and (e) resonance hybridization.*

complete the electron configuration of all the atoms (*Fig. 1b*). It is customary to replace the bonding pairs of shared electrons with one line for each pair – each line representing a bond – with multiple lines representing multiple bonds (*Fig. 1c*).

Although main group elements tend to adopt inert gas configurations, which may be represented by eight valence electrons (an **octet**), or two in the case of helium, a number of elements are energetically stable with **incomplete octets**. The most commonly cited example is boron, which is stable with six valence electrons as in $BF_3$, or Be with four as in $BeCl_2$. Larger elements are capable of **hypervalency**, where it is energetically favorable for more than eight valence electrons to be held in an expanded octet. Examples of this are $PF_5$ (ten valence electrons) and $XeF_4$ (twelve valence electrons) (*Fig. 1d*).

In many compounds, it is possible to devise two or more equivalent bonding schemes (**canonical forms**) by varying only the position of multiple bonds. Neither structure adequately describes the bonding by itself, and the molecule is represented as a combination of the two, known as a **resonance hybrid**, by a double-headed arrow (*Fig. 1e*).

**VSEPR theory**

Valence shell electron pair repulsion (VSEPR) theory is an elementary approach to explaining the shapes of molecules. The theory treats each atom in a molecule in isolation, and describes the geometry of the bonds and non-bonding electron pairs around it. The basic assumption is that the electron pairs around an atom, both bonding and non-bonding, will adopt a geometry which will minimize repulsive forces by maximizing the distances between pairs.

The precise geometry of electron pairs around a central atom depends firstly upon the number of electron pairs which are present. For certain numbers of electron pairs (2, 3, 4, 6), it is possible to adopt a geometry in which the pairs are equidistant. For atoms with 5 or 7 electron pairs, this is not possible, and the maximum separation involves some compromise (*Table 1*).

*Table 1.  The dependence of molecular geometry on the number of electron pairs*

| No. of electron pairs | Geometry | Interbond angles | Arrangement | Example |
|---|---|---|---|---|
| 2 | Linear | 180° | | $BeCl_2$ |
| 3 | Trigonal planar | 120° | | $BF_3$ |
| 4 | Tetrahedral | 109.5° | | $CH_4$ |
| 5 | Trigonal bipyramid | 120° and 90° | | $PF_5$ |
| 6 | Octahedral | 90° | | $SF_6$ |
| 7 | Pentagonal bipyramid | 90° and 72° | | $IF_7$ |

The basic geometry is modified by the variations in repulsion strengths between the electron pairs. Because the charge in bonding pairs is somewhat offset by the presence of the bonded nuclei, the repulsion increases in the order:

bonding pair:bonding pair < non-bonding pair:bonding pair < non-bonding-pair:non-bonding pair

In determining the geometry of neighboring atoms around a central atom, the number of electron pairs is ascertained, so giving the underlying geometry. The pairs are then arranged so as to give the maximum distance between non-bonding pairs, giving the actual geometry of the neighboring bonded atoms. In $CH_4$, the bonding electron pairs adopt a perfect tetrahedral arrangement (*Fig. 2a*). In ammonia, $NH_3$, whilst there are four valence shell pairs, giving an underlying tetrahedral geometry, the greater repulsive effect of the non-bonding pair forces the bonding pairs closer to one another than in the ideal tetrahedral geometry (*Fig. 2b*). For a tetrahedral arrangement of four bonding pairs, the bond angle is 109.5°. With one non-bonding electron pair, this reduces to 107.3°, and with two non-bonding electron pairs (e.g. water, $H_2O$), the angle is further reduced to 104.5° (*Fig. 2c*).

A multiple bond, representing two or more electron pairs, is treated as a single electron pair, but with a greater electron density, and so has a greater electron repulsive effect than a non-bonding pair.

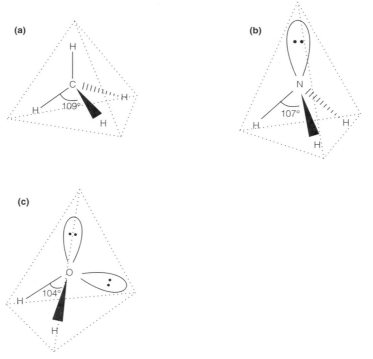

*Fig. 2.    VSEPR and the geometric arrangement of bonds in molecules with non-bonding electrons.*

# H2 VALENCE BOND THEORY

## Key Notes

**Valence bond theory of diatomic molecules**

Valence bond theory focuses attention on formation of the individual bonds in molecules. Bonds are generated from combinations of the atomic orbitals from each of the bonding atoms, and the bond is mathematically described as a function of these atomic orbitals. Where the resulting bond has cylindrical symmetry about the bond axis, it is termed a σ (sigma) bond. Elements with accessible $p$ orbitals may generate bonds by sidelong overlap of the orbitals to give a π (pi) bond.

**Valency**

The valency of many elements is greater than that predicted from the number of unpaired atomic electrons. Promotion – raising one electron from a pair into a higher energy orbital – creates two additional unpaired electrons in an atom, and so increases the valency by two. The energy required for electron promotion is offset by the energy recouped in forming two additional chemical bonds.

**Polyatomic molecules**

The shapes of polyatomic molecules are poorly described by pure atomic orbitals. Deviation of actual bond angles from the angles between pure atomic orbitals is accounted for by hybridization.

**Hybridization**

Atomic orbitals are combined into hybrid orbitals, whose shape is defined by the geometry and proportion of the atomic orbitals. $sp^3$ hybrid orbitals are tetrahedrally arranged, $sp^2$ hybrid orbitals are arranged in a trigonal planar fashion, and $sp$ orbitals are linearly arranged. More complex geometric configurations may be obtained by hybridization involving $d$ and $f$ orbitals.

**Related topics**

The wave nature of matter (G4)
The structure of the hydrogen
  atom (G5)
Many-electron atoms (G6)

Elementary valence theory (H1)
Molecular orbital theory of diatomic
  molecules I (H3)

**Valence bond theory of diatomic molecules**

**Valence bond theory** is a **quantum-mechanical** description of molecular bonding which focuses attention on formation of the bond itself. As for Lewis theory (see Topic H1), bonds are generated by pairing up electrons on one atom with electrons on a second atom, but the nature of the atomic orbitals themselves are also considered. Electrons are paired in the course of bond formation and the spin of the individual electrons must be taken into account. **Spin pairing** is a requirement that the electrons from one atom are paired only with an electron of opposite spin. This ensures that a molecular bond is created in which the two electrons do not occupy the same quantum state, and so comply with the **Pauli exclusion principle** (see Topic G6).

Bonds are formed from combinations of the atomic orbitals from each of the bonding atoms, and the mathematical description of the molecular bond is therefore a function of these atomic orbitals. It is a fundamental requirement that the electrons in the molecular orbital are indistinguishable, and the simplest orbital function compatible with this is the **Heitler-London wavefunction**:

$$\psi = \psi_A(1)\psi_B(2) + \psi_A(2)\psi_B(1)$$

The first product describes the case of electron 1 in orbital A ($\psi_A(1)$) and that of electron 2 in orbital B ($\psi_B(2)$), with the second product describing the complementary situation. The two terms are not identical, as the electrons possess opposite spins. The resulting wavefunction describes the condition where either electron may be found on either of the bonded atoms.

In the simplest example, that of a hydrogen molecule, the atomic 1s orbitals are the sole contributors to the bond, and the wavefunction takes the form:

$$\psi_{H-H} = \psi_{H_{1s}A}(1)\psi_{H_{1s}B}(2) + \psi_{H_{1s}A}(2)\psi_{H_{1s}B}(1)$$

The physical results of this mathematical expression are illustrated in *Fig. 1a* and *1b*. The resulting bond has cylindrical symmetry about the bond axis, and is termed a **σ (sigma) bond**.

In elements with accessible *p* orbitals, such as oxygen or nitrogen, more complex bonding may be obtained. The two atomic *p* orbitals which are parallel to the bonding axis (the $p_z$ orbitals, by convention) may be combined so as to form a σ bond (*Fig. 1c*), but it is also possible for *p* orbital pairs which are perpendicular to the bonding axis ($p_x$ on A and B or $p_y$ on A and B) to combine to give **π (pi) bonds** (*Fig. 1d*). The strength of the π-bond is significantly less than

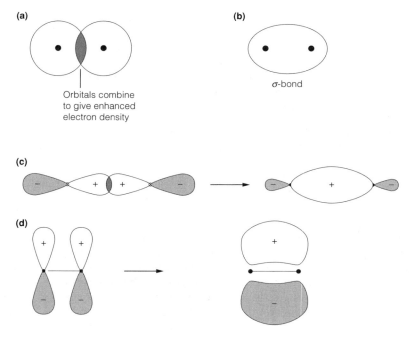

*Fig. 1.* (a) Two free hydrogen atoms and (b) the σ-bond formed from the combination of their 1s orbitals. (c) A σ-bond generated from two $p_z$ orbitals. (d) A π-bond generated from two $p_y$ orbitals.

that of the σ-bond, as the 'sidelong' overlap of the $p$ orbitals is less than that of the 'direct' overlap (the products $\psi_{p_xA}(1)\psi_{p_xB}(2)$ and $\psi_{p_xA}(2)\psi_{p_xB}(1)$ being correspondingly reduced). Each pair of atomic $p$ orbitals forms one molecular π-bond, giving a maximum of three molecular bonds from each set of $p$ orbitals – one σ-bond and two mutually orthogonal π-bonds. The π-bonds do not have cylindrical symmetry, having instead a nodal plane parallel with the bonding axis.

**Valency**

The simple interpretation of the valence bond model fails to account for the **valency** (number of bonds) or multiple valency states of many elements. The valence electron configuration of silicon, $3s^23p_x{}^13p_y{}^1$, for example, suggests a valency of two arising from the two singly occupied $p$ orbitals. The valency in fact increases through **promotion**, i.e. raising an electron into a higher energy orbital. This process breaks up an electron pair to give two additional unpaired electrons, and so increases the valency by two. In the case of silicon, promotion of an electron from the $3s$ to the $3p$ orbital results in a tetravalent configuration of $3s^13p_x{}^13p_y{}^13p_z{}^1$. In some elements, the process may occur several times, each time increasing the valency by two. In all cases, the energy required for promotion of the electron must be offset by the energy recouped in forming two additional chemical bonds for this process to occur.

**Polyatomic molecules**

The principle of spin pairing of electrons in singly occupied orbitals to form bonds may be extended to molecules with any number of atoms, with the available atomic orbitals on one atom combining with those on two or more other atoms.

The **valence bond** approach is broadly successful in predicting the number of available bonds, but is very unsatisfactory in its ability to predict the shape of molecules. In a commonly used example, the basic theory predicts that the bonding in water, $H_2O$, would consist of two σ-bonds formed from pairing of electrons in the hydrogen $1s^1$ orbitals and two oxygen $p$ orbitals. As the atomic $p$ orbitals are orthogonal, valence bond theory predicts that the resulting σ-bonds are at 90° to one another. In fact, the inter-bond angle is closer to 104°. The deviation of actual bond angles from the angles between pure atomic orbitals is accounted for by **hybridization.**

**Hybridization**

**Hybridization** is the process of combining pure atomic orbitals so as to circumvent the rigid geometry which the pure orbitals require. In this way, valence bond theory becomes far more able to account for molecular shapes. The pure orbital functions have both negative and positive signs. By directly combining the atomic orbitals, these negative and positive regions are added so as to enhance the amplitude of the resulting orbitals in some directions, and to diminish their amplitude in others. The resulting combinations of pure orbitals are termed **hybrid orbitals.**

The most significant application of hybridization is in the shapes of the molecules involving the elements nitrogen, oxygen and particularly carbon. Combinations of one $s$ and one $p$ orbital give rise to two $sp$ **hybrid** orbital combinations (*Fig. 2a*). For a trigonal planar geometry, two $p$ orbitals combine with one $s$ orbital to yield three $sp^2$ **hybrid orbitals** (*Fig. 2b*) and for a tetrahedral geometry, a combination of one $s$ orbital and three $p$ orbitals is used, gives rise to four $sp^3$ **hybrid orbitals** (*Fig. 2c*).

The $sp$, $sp^2$, and $sp^3$ hybrids represent limiting hybrids, and it is possible to combine the orbitals in such a way as to optimize the valence bonds to the required geometry. More complex geometric configurations may be obtained by

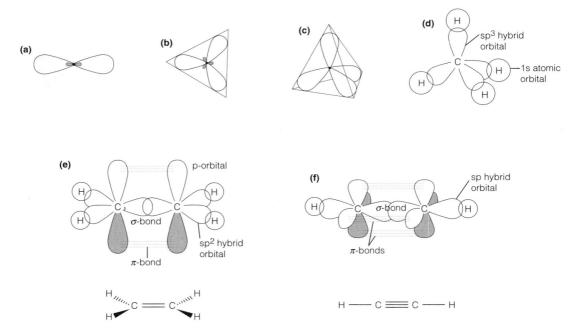

Fig. 2.    (a) Two sp hybrid orbitals with a linear arrangement. (b) Three sp² hybrid orbitals arranged in a trigonal planar geometry. (c) Four tetrahedrally arranged sp³ hybrid orbitals. (d) The valence bond model of methane, showing the bonding between carbon sp³ orbitals and four hydrogen 1s orbitals. (e) The sp² σ bonding and pure p orbital π bonding interaction in ethene, showing the origin of the double bond. (f) The sp σ bonding and two pure p orbital π bonding interactions in ethyne, showing the origin of the triple bond.

hybridization involving $d$ and $f$ orbitals, so that an $sp^3d$ hybrid orbital set generates a trigonal bipyramidal geometry, and an $sp^3d^2$ hybrid orbital set is octahedrally arranged.

The hybridization in an atom is imposed as a result of the molecular environment, and its precise nature is determined by the most effective way in which the free energy can be reduced through bonding. Hybridization is *not* a property of a free atom and does not occur prior to the bonding process, but takes place in parallel with bond formation. In practical bonding applications, however, the hybrid orbitals behave identically to pure atomic orbitals, and may be treated likewise. Hence the spatial arrangement of the bonds of methane, for example, may be accurately reproduced (within the constraints of the theory) from the four $sp^3$ hybrid orbitals of carbon, and the four 1s atomic orbitals of hydrogen (*Fig. 2d*).

Hybrid orbitals are invoked in order to enable formation of σ bonds, whose geometry in turn defines the geometry of the molecule. In the case of carbon $sp$ and $sp^2$ hybrids, for example, this leaves one and two unused $p$ orbitals, respectively. In both cases, the remaining pure $p$ orbitals allow formation of multiple bonds with neighboring atoms via π bonds (*Fig. 2e* and $f$).

# H3 MOLECULAR ORBITAL THEORY OF DIATOMIC MOLECULES I

## Key Notes

**The Born-Oppenheimer approximation**

Direct calculation of the molecular orbitals is possible only in the case of $H_2^+$. For more complex molecules, a range of approximations must be made in order to predict the nature of the molecular orbitals. In all cases, it is necessary to assume that the nuclei are stationary relative to the motion of the electrons. This is known as the Born-Oppenheimer approximation, and allows the internuclear geometry and motion to be treated completely separately from that of the electrons.

**The LCAO approximation**

In all cases other than that of $H_2^+$, the molecular orbitals may be approximately derived from a linear combination of atomic orbitals. The atomic wavefunctions are combined in a linear combination to give a molecular orbital, $\psi = \sum_m c_m \psi(m)$, where $c_m$ is the mixing coefficient. $c_m$ may be varied for all the atomic orbitals so as to minimize the energy of the resulting molecular orbital. The variation principle states that the lowest energy calculated orbital most accurately describes the actual molecular wavefunction.

**Bonding and antibonding orbitals**

A molecular bonding orbital differs from the atomic orbitals from which it is derived as it increases the probability of finding an electron in the internuclear region. This reduces the free energy of the electrons, and that of the molecule as a whole. Antibonding molecular orbitals are derived by subtraction of one or more atomic wavefunction from the others. The energy of the resulting antibonding molecular orbital is greater than that of the atomic orbitals, since a node in the internuclear electron density causes an increase in the internuclear repulsion.

**Potential energy curves**

The stability of a molecule is heavily dependent upon the extent to which the orbitals are allowed to overlap. The extent of the orbital overlap is determined by the internuclear distance. At relatively large separations, the energy decreases as the nuclei are brought together, and the atomic orbital overlap increases, whereas at low internuclear distances a repulsion term is dominant. There is a point at which these two opposing effects balance, and the molecule adopts its lowest free energy state. For the antibonding orbital, there is a fully repulsive interaction between the nuclei at any distance.

**Related topics**

The wave nature of matter (G4)
The structure of the hydrogen
   atom (G5)
Many-electron atoms (G6)

Elementary valence theory (H1)
Valence bond theory (H2)
Molecular orbital theory of diatomic
   molecules II (H4)

**The Born-Oppenheimer approximation**

In contrast to valence bond theory (see Topic H2), **molecular orbital theory** attempts to describe the bonding orbitals in a molecule in their entirety, as opposed to focusing on the creation of each bond individually. Just as the explicit calculation of atomic orbitals is only possible for the hydrogen atom, direct calculation of the molecular orbitals is possible only in the simplest possible molecule, $H_2^+$. Unlike atomic wavefunctions, the molecular wavefunction must describe the relative motion of nuclei in addition to the motion of the electrons. Primarily because of their relative masses, the electrons move some $10^3$ times faster than the nucleus, and the **Born-Oppenheimer approximation** simplifies the calculation of molecular orbitals by assuming that the nuclei are stationary relative to the motion of the electron. This approximation allows the internuclear repulsion terms to be treated completely separately from the electrostatic behavior of the electrons.

The potential energy of an electron in the electric field resulting from two protons is readily calculated using:

$$V = \frac{-e^2}{4\pi\varepsilon_o}\left(\frac{1}{r_{H1}} + \frac{1}{r_{H2}}\right)$$

where $r_{H1}$ and $r_{H2}$ are the distances of the electron from each proton. Incorporation of this potential energy expression into the Schrödinger equation (see Topic G4) yields the exact solutions for the hydrogen molecular ion. As with atomic orbitals, the impossibility of calculating orbitals for systems with three or more bodies in relative motion makes mathematical solutions for the molecular orbitals impossible. Molecular orbital theory therefore makes the approximation that the molecular orbitals may be formed by the **linear combination of atomic orbitals (LCAO)**.

**The LCAO approximation**

The total wavefunction for a molecule is given by:

$$\psi = \prod_n \psi_n$$

where $\psi_n$ represents the wavefunction for each electron in the molecule.

Explicit calculations for the molecular orbitals $H_2^+$ show that the lowest energy solution of the Schrödinger equation is given by the addition of the two $1s$ orbitals. In all other cases, it is necessary to make the approximation that molecular orbitals may be calculated from a linear combination of atomic orbitals. This linear combination generates molecular orbitals by direct addition of atomic orbitals on the bonding atoms. The wavefunctions for two electrons, $\psi(A)$ and $\psi(B)$, in the atomic orbitals on two atoms, A and B respectively, are combined to form a **molecular orbital**. The molecular orbital wavefunction is given by:

$$\psi(MO) = c_A\psi(A) + c_B\psi(B) \qquad \text{or} \qquad \psi = \sum_m c_m\psi(m) \qquad \text{(general case)}$$

where $c_m$ is a **mixing coefficient**, which is calculated for each orbital so as to minimize the molecular energy as calculated using the Schrödinger equation. It is a further requirement that the mixing coefficients should be **normalized**, so that $\sum_m c_m^2 = 1$. When the bond is formed between two identical nuclei there can be no distinction between the nuclei, and so the mixing coefficients (and therefore the orbital contributions), are equal. The resulting molecular orbital becomes more accurately described as more of the available atomic orbitals are included in the calculation. The **variation principle** states that the lower the energy of the calculated orbital, the more accurately it describes the actual molecular wavefunction.

**Bonding and antibonding orbitals**

The probability, $p$, of an electron in an orbital, $\psi$, being in an infinitesimal volume $d\tau$ at a point $r$ is obtained through the **Born interpretation**:

$$P(r) = \psi(r)^2 d\tau \quad \text{(note that } \int_{r=0}^{\infty} \psi(r)^2 d\tau = 1 \text{, as the electron must exist somewhere)}$$

For $H_2^+$, the molecular orbital **probability function** is then given by:

$$P(r) = (\psi(A) + \psi(B))^2 d\tau = \psi(A)^2 d\tau + \psi(B)^2 d\tau + 2\psi(A)\psi(B)d\tau$$

This differs from the atomic orbitals as it represents the probability of finding an electron in the two constituent atomic orbitals plus an additional term. This term has the effect of increasing electron density in the internuclear region, and decreasing it in other regions (*Fig. 1a*). Since electrons are waves, the enhanced electron density can be rightly compared to the constructive interference of two waves. The enhanced electron density between the nuclei enables the electrons to associate strongly with both nuclei simultaneously, thereby reducing the free energy of the electrons, and so the molecule as a whole.

The structure of a molecule depends upon the formation of bonds holding the nuclei in their relative positions, and these result from occupied **bonding orbitals** (*Fig. 1b*). Molecular bonding orbitals are derived from linear combination of atomic orbitals by the addition of the component atomic orbitals, and act to decrease the free energy of the molecule. The explicitly calculated result for $H_2^+$ reveals that the next highest energy solution to the Schrödinger equation is an **antibonding orbital** corresponding to the subtraction of one wavefunction from the other:

$$\psi(MO) = \psi(A) - \psi(B)$$

As with the bonding orbital, the probability function for this molecular orbital is given by:

$$P(r) = (\psi(A) - \psi(B))^2 d\tau = \psi(A)^2 d\tau + \psi(B)^2 d\tau - 2\psi(A)\psi(B)d\tau$$

The negative term now corresponds to a decrease of electron density in the internuclear volume of the molecule. The electron density decrease raises the energy of the molecule above that of the free atoms. There is now a **node** – a line of zero electron density – in the molecular orbital, between the two atoms (*Fig. 1c*).

It is usual to denote an antibonding orbital with an asterisk to distinguish it from the bonding orbital. In the case, therefore, of two $s$ orbitals interacting to form both a bonding and an antibonding orbital, both molecular orbitals will be $\sigma$ orbitals. The antibonding orbital is denoted as $\sigma^*$, allowing its immediate distinction from the bonding orbital, $\sigma$.

**Potential energy curves**

The degree of stabilization conferred on a molecule by the overlap of atomic orbitals is heavily dependent upon the extent to which the orbitals are allowed to overlap. The extent of the orbital overlap is in turn determined by the internuclear distance.

A potential energy curve may be plotted for a molecule, and is constructed by plotting the energy of the molecule as a function of internuclear distance. For the bonding orbital at relatively large separations, the energy decreases as the nuclei are brought together, and the atomic orbital overlap increases. The total wavefunction for the molecule also includes an internuclear repulsion term, which increases with increasing nuclear proximity, and the energy of the

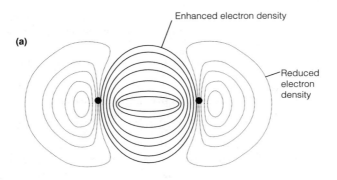

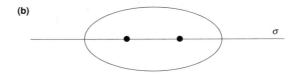

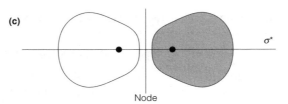

Fig. 1.    (a) Electron density difference between atomic 1s and molecular σ bonds in hydrogen; (b) hydrogen σ molecular bonding orbital; (c) hydrogen σ* antibonding molecular orbital.

molecule increases rapidly at smaller internuclear distances. At an intermediate internuclear distance, there is a balance between these two opposing effects, and the molecule adopts its lowest free energy state (*Fig. 2*). For the antibonding orbital, the decrease in electron density between the nuclei means that there is a fully repulsive interaction between the nuclei at any distance, and this increases rapidly as the internuclear separation decreases.

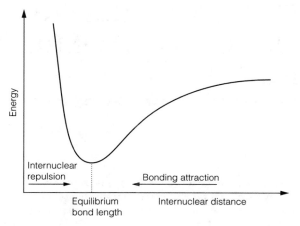

*Fig. 2.* *Potential energy curve for a bonding molecular orbital in a diatomic molecule.*

# H4 MOLECULAR ORBITAL THEORY OF DIATOMIC MOLECULES II

## Key Notes

**Hydrogen and helium molecules**

Molecular orbital energy levels may be represented by a molecular orbital energy level diagram, which illustrates the relative energy of the molecular and atomic orbitals, and their relationship. The bonding molecular orbital is of lower energy than that of the separate atomic orbitals, and the converse is true for the antibonding orbital. The available electrons are placed in pairs into the molecular orbitals, with the lowest energy molecular orbitals being filled first. In the hydrogen molecule, both electrons occupy the bonding molecular orbital, giving an overall bonding interaction. In $He_2$, two electrons occupy each of the $\sigma$ and $\sigma^*$ orbitals, giving no net bonding.

**Homonuclear second row diatomic molecules**

In second row diatomic molecules, linear combinations of the $p$ orbitals give both a $\sigma$ orbital and two degenerate $\pi$ orbitals. As the molecular orbitals are qualitatively unchanged for the second row diatomic molecules, the same molecular orbital diagram may be used by entering the correct number of electrons. Detailed analysis reveals that the highest occupied $\pi$ and $\sigma$ orbitals exchange positions between the elements nitrogen and oxygen. Core orbitals do not make a significant contribution to the bonding, as each pair of bonding and antibonding orbitals is fully occupied, leaving no net bonding contribution.

**Parity**

Molecular orbitals in homonuclear molecules may be described in terms of their symmetry with respect to a point of inversion at the center of the bond. A molecular orbital whose sign is unchanged by inversion is termed gerade, $g$, and one whose sign is inverted is ungerade, $u$.

**Term symbols**

The total spin angular momentum quantum number, $\Sigma$, describes the overall spin of the electrons in the molecule, and is quoted as its multiplicity, $(2\Sigma+1)$. The orbital angular momentum is described by a quantum number $\Lambda$. For an electron in a $\sigma$ orbital, $\Lambda = 0$, for an electron in a $\pi$ orbital, $\Lambda = \pm1$, etc. The term symbol representing a specific electron configuration is written in the form $^{2\Sigma+1}\Lambda_{parity}$. The ground state of nitrogen is written as $^1\Sigma_g$, for example.

**Heteronuclear diatomic molecules**

Molecular orbitals in heteronuclear diatomic molecules do not have equal mixing coefficients for corresponding atomic orbitals, leading to unequal orbital distributions over the two nuclei. Electrons spend more time around one atom than the other, generating a dipole over the length of the bond. This leads to a polar covalent bond. The element which most strongly attracts the electrons is referred to as the more electronegative element, and the strength of the attraction is most commonly measured using the Pauling electronegativity scale.

| **Dipole moments** | A positive charge, $+q$ and negative charge, $-q$, separated by a distance $R$, give rise to an electric dipole moment, a vector directed from the positive to the negative charge across the molecule, with magnitude $qR$. The value of the dipole moment is generally reported in debye, D, where 1 debye is equal to $3.336 \times 10^{-30}$ C m. The calculation of dipole moments in polyatomic molecules may be calculated by vector addition of the dipole moments. |
|---|---|
| **Related topics** | The wave nature of matter (G4)  Elementary valence theory (H1)  The structure of the hydrogen  Valence bond theory (H2)  atom (G5)  Molecular orbital theory of diatomic  Many-electron atoms (G6)  molecules I (H3) |

**Hydrogen and helium molecules**

The energy levels of diatomic molecules are conventionally represented in the form of a **molecular orbital energy level diagram.** The orbital is represented by a horizontal line whose vertical position indicates the relative energy of that orbital. The atomic orbital energy levels of the two constituent atoms are arranged either side of the molecular orbital energy levels, usually with lines linking related molecular and atomic orbitals. This is illustrated in *Fig. 1* for the hydrogen molecule.

The **molecular bonding orbital** for hydrogen is generated by the linear combination of the atomic 1s orbitals, and as there can be no distinction between the 1s orbitals from each atom, the mixing coefficients (see Topic H3) for the orbitals are equal. This means that the molecular orbital is composed of equal proportions of each 1s orbital, and an electron in an orbital of this nature therefore spends equal time around each nucleus.

The bonding molecular orbital is of lower energy than that of the separate atomic orbitals, and is therefore placed below this level on the molecular orbital diagram. The converse is true for the antibonding orbital, which is of higher energy than both the bonding orbital, and the isolated atomic orbitals.

As with atomic orbitals, it is possible to place a maximum of two electrons of opposite spin into each molecular orbital, and the total number of electrons occupying the molecular orbitals is equal to the number of electrons in the isolated species. These electrons are placed in pairs into the molecular orbitals, with the lowest energy molecular orbitals being filled first.

In the hydrogen molecule, both electrons occupy the bonding molecular orbital, and this bonding wavefunction dominates the molecule. Promotion of an electron into the $\sigma^*$ orbital, however, creates one antibonding and one

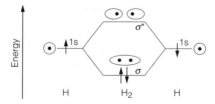

*Fig. 1.   The molecular orbital energy diagram for hydrogen, showing the relative positions of the atomic and molecular orbital energies (not to scale), the schematic geometry of the orbitals, and the electron configuration. Electrons are indicated by vertical arrows.*

bonding electron, which leaves no net bonding, and this excited form of the hydrogen molecule is unstable. The next simplest diatomic molecule which could be formed is $He_2$. Two electrons are donated from each atom, and so two electrons occupy each of the σ and σ* orbitals, giving no net bonding (*Fig. 2*). $He_2$ is consequently such an unstable molecule that it does not exist.

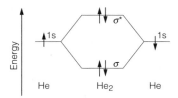

*Fig. 2.   The hypothetical helium molecule, showing equal population of bonding and anti-bonding interactions.*

**Homonuclear second row diatomic molecules**

Diatomic molecules composed of identical atoms $N_2$ and $O_2$, for example, are referred to as **homonuclear diatomic molecules**. In the second and subsequent rows of the periodic table, the bonding interactions involve both $p$ and $s$ orbital interactions. By convention, the bonding axis is taken to be the $z$ axis and the linear combinations of the $p_z$ orbitals differ from those of the $p_x$ and $p_y$ orbitals.

The lobes of the atomic $p_z$ orbitals interact directly and relatively strongly along the bonding axis to form molecular bonding and antibonding σ orbital combinations. The linear combinations of the $p_x$ and $p_y$ orbitals on the other hand give rise to two bonding and antibonding π orbital combinations (*Fig. 3*). The overlap of the $p_x$ and of the $p_y$ orbitals is smaller than that of the $p_z$ orbitals, and this is reflected in the energy of the σ bond which is lower than that of the two π bonds. The two π bonds are of equal energy and are said to be **degenerate**.

**Core atomic orbitals** do not make a significant contribution to the bonding in a molecule for two reasons. Firstly, there is no significant overlap of these orbitals, and the energy of the bonding orbitals is not significantly lower than the atomic orbitals, and secondly, each pair of bonding and antibonding is fully occupied, leaving no net bonding contribution.

The molecular orbital diagram for a second row diatomic, $O_2$, is shown in *Fig. 3*, with eight 2p electrons occupying the two degenerate bonding π-orbitals, the bonding σ-orbital and two degenerate π* orbitals to give a double bond overall.

The σ bonding and antibonding molecular orbitals derived from the 1s and 2s atomic orbitals are fully occupied and so have no net bonding effect. In oxygen, the two highest energy electrons are placed separately into the $π_x$* and $π_y$* orbitals to give two unpaired electrons, which also confer oxygen with significant paramagnetism.

As the molecular orbitals are qualitatively unchanged, the same molecular orbital diagram may be used to describe any second row diatomic molecule or ion, by simply entering the correct number of electrons. However, detailed analysis reveals the importance of including all the atomic orbitals when generating the linear combination of orbitals. The narrow energy gap between the 2s and 2p orbitals in the early part of the second row leads to mixing of the 2s atomic orbitals with the $2p_z$ orbitals. This raises the energy of the $2p_z$ generated σ orbitals, and lowers the energy of the 2s generated σ* orbital, so that the highest occupied π and σ bonding orbitals exchange positions between nitrogen and oxygen (*Fig. 4*).

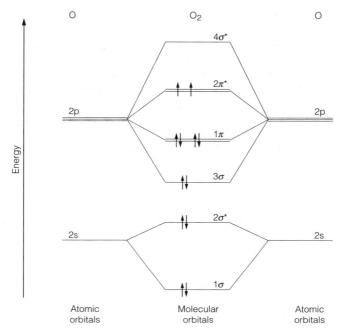

*Fig. 3.    Molecular orbital diagram for $O_2$ (valence shell electrons only).*

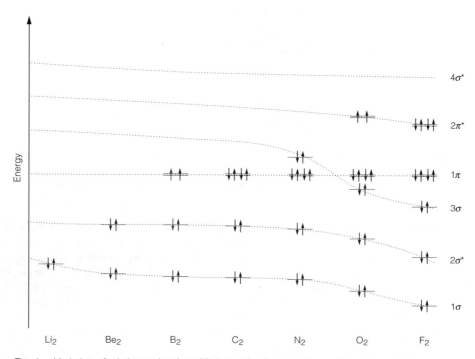

*Fig. 4.    Variation of relative molecular orbital energies for second row diatomic molecules.*

**Parity**

A **center of symmetry** is located at the midpoint of the bond in a homonuclear diatomic molecule, i.e. any point on the molecule may be reflected through the center of symmetry to give a second point with identical physical properties. Molecular orbitals in homonuclear molecules may be described in terms of their symmetry with respect to this point, since, although the electron density is identical by reflection through the center of symmetry, the sign of the wavefunction need not be. A molecular orbital whose sign is unchanged by reflection is termed **gerade**, abbreviated to **g**, whilst one whose sign is inverted is **ungerade, u**. The sign of a bonding $\pi$ orbital is inverted through the center of symmetry, and is denoted $\pi_u$. The sign of a bonding $\sigma$ orbital is not inverted and is denoted $\sigma_g$ (*Fig. 5*).

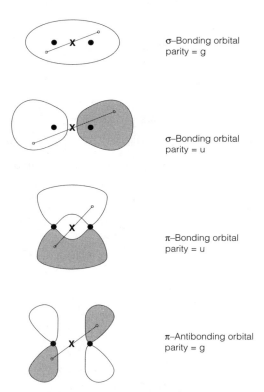

$\sigma$–Bonding orbital
parity = g

$\sigma$–Bonding orbital
parity = u

$\pi$–Bonding orbital
parity = u

$\pi$–Antibonding orbital
parity = g

*Fig. 5. Parity relationships for some molecular orbitals. The center of inversion is denoted by X.*

**Term symbols**

As with the distributions of electrons in atomic orbitals, the distribution of electrons in molecular orbitals may be denoted with the use of a **term symbol**. The total **spin angular momentum quantum number**, $\Sigma$, describes the overall spin of the electrons in the molecule, and is quoted as its **multiplicity**, $(2\Sigma+1)$. A **closed shell** has $\Sigma = 0$, and a multiplicity of 1 (**a singlet**), a single unpaired electron has $\Sigma = \frac{1}{2}$ and multiplicity 2 (**a doublet**) two unpaired electrons have $\Sigma=1$ and multiplicity 3 (**a triplet**).

The **orbital angular momentum** is described by the quantum number $\Lambda$. For an electron in a $\sigma$ orbital, $\Lambda=0$, for a $\pi$ orbital, $\Lambda=\pm1$, etc. Pairs of degenerate $\pi$ orbitals contribute no angular momentum, as their contributions cancel out.

The molecular term symbol is created as for an atomic term symbol. The value of $\Lambda$ is denoted by $\Sigma$ for $\Lambda = 0$ (note that this label is not related to the symbol for the spin angular momentum), $\Pi$ for $\Lambda = 1$, $\Delta$ for $\Lambda = 2$, etc. (c.f. S,P,D... for atomic term symbols), and the multiplicity of the spin angular momentum is added to this as a superscipt (e.g. $^1\Pi$, $^3\Delta$). The parity of the overall wavefunction may be added as a subscript. The ground state term symbol for nitrogen is $^1\Sigma_g$, and that of oxygen is $^3\Sigma_g$, for example.

**Heteronuclear diatomic molecules**

Diatomic molecules composed of two different elements, such as CO or NO are termed **heteronuclear**. Molecular orbitals in heteronuclear diatomic molecules are constructed in the same manner as those in homonuclear molecules. The bonding differs from that of a homonuclear diatomic molecule in the form of the LCAO:

$$\psi(MO) = c_1\psi_1 + c_2\psi_2$$

For a molecular orbital in a homonuclear diatomic molecule, the mixing coefficients, $c_1$ and $c_2$ (see Topic H3) are equal whereas the **mixing coefficients** for corresponding atomic orbitals are no longer equal in a heteronuclear species.

As a result of the inequality of the mixing coefficients, all the molecular orbitals have, to varying degrees, unequal distributions over the two nuclei. Electrons therefore spend more time around one atom than the other, on average, and this gives rise to a **dipole** over the length of the bond. The greater the energy difference between the corresponding atomic orbitals, the greater the difference between the mixing coefficients, and the greater the localization of electrons around one of the atoms. This is illustrated schematically for the carbon monoxide molecule in *Fig. 6*. In the highest occupied $\pi$ bonding orbital, the mixing coefficient for the oxygen $2p$ orbitals is greater than that of the carbon $2p$ orbitals, so giving the $\pi$ orbital a higher degree of oxygen $2p$ character than

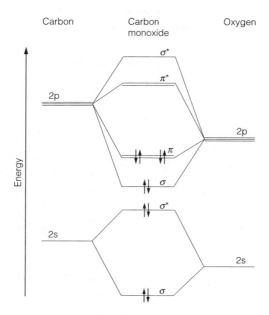

*Fig. 6.  Molecular orbital diagram for a heteronuclear diatomic molecule (CO).*

carbon $2p$ character. The net effect is for the electrons to spend more time around the oxygen than the carbon. As more electronic charge accumulates on one atom over the other, this leads to a **polar covalent bond**. An extreme example of this may be seen as the basis of the ionic bonding in, for example, KCl.

The element which most strongly attracts the electrons is referred to as the more **electronegative** element. Conversely, the element which holds them less strongly is referred to as being more **electropositive.** Among the main group elements, the more electronegative elements tend to be in the later groups and earlier periods of the periodic table (top right), whilst the electropositive elements tend to be located in the early groups and later periods (bottom left). The strength of the attraction is most commonly measured using the **Pauling electronegativity scale**, with values, denoted $\chi$, ranging between 4 for fluorine, the most electronegative element, down to *ca*. 0.6 for francium, the most electropositive.

**Dipole moments**

A **polar covalent bond** implies that one atom in a molecule will be more positively charged than the other. The positive charge, $+q$ and the negative charge, $-q$, separated by a distance $R$, give rise to an **electric dipole moment**, $\mu$. This is a vector directed from the positive to the negative charge across the molecule, with magnitude $qR$. This vector is usually represented by an arrow directed from the positive to the negative charge, with a positive sign included to indicate the positive end, thus: $\longmapsto$.

In order to generate convenient values, the dipole moment is generally reported in **debye**, **D**, where 1 debye is equal to $3.336 \times 10^{-30}$ C m. Water, for example, has a dipole moment of 1.85 D. The size of the dipole moment in debyes between two atoms, A and B may often be estimated from their respective **Pauling electronegativities**, $\chi(A)$ and $\chi(B)$:

$$\mu \approx \chi(A) - \chi(B)$$

The dipole moments in polyatomic molecules may be calculated by vector addition of the dipole moments for each band (*Fig. 7*).

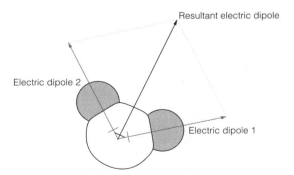

*Fig. 7.   Overall molecular dipole moments derived by vector addition of individual dipole moments in some polyatomic molecules.*

# H5 STRONG SOLID PHASE INTERACTIONS

## Key Notes

**Bonding in solids: band theory**

Molecular orbitals in large, extended solids form energy bands, rather than discrete energy levels, and an electron may hold any of a continuous range of energies. Two complementary theories form the basis of band theory: the tightly bound electron model (also known as the tight binding approximation) and the free electron model.

**Tightly bound electron model**

The tightly bound electron model extends the LCAO approach to the overlapping of very large numbers of molecular orbitals. This results in large numbers of molecular orbitals with the fully bonding orbital at the lowest energy, and the fully antibonding orbital at the highest. The energy difference between the upper and lower orbitals is known as the band width, and the number of molecular states per unit binding energy is referred to as the density of states. In real solids bands are formed in three dimensions using $s$, $p$, $d$, and $f$ orbitals and both the structure of the bands and the density of states diagrams may be extremely complex.

**Free electron model**

The free electron model reduces the problem of the energy levels to that of a particle in a box. The maximum energy of an electron, mass $m_e$, in this model is given by:

$$E_{max} = \frac{h^2}{2m_e(3\rho/8\pi)^{\frac{2}{3}}}$$

where $\rho$ is the density of the electrons. Bands in the free electron model are open-ended, and do not have an upper energy limit. More accurate approaches to band theory account for both the highly delocalized nature of the electrons, and the regular array of nuclear potentials.

**Electrons in bands**

As with molecular orbitals, the energy levels in a band are progressively filled from the lowest energy upwards. The Fermi level, $E_f$, is defined as the energy of the orbital for which the probability of electron occupation is ½. The distribution of electrons is described by Fermi-Dirac statistics. The probability $f(E)$ of an electron occupying an energy $E$ at a temperature, $T$ is given by:

$$f(E) \propto \frac{1}{1 + \exp\left[(E - E_f)/k_B T\right]}$$

**Metals, insulators and semiconductors**

In a metal, at least one band is partially filled, caused either by a deficit of electrons or by band overlap. In an insulator, the highest occupied band is filled, and there is a significant energy gap ('band gap') between this and the lowest unoccupied band. In a semiconductor, thermal excitation across the band gap is possible at ambient temperatures which enables conduction to take place.

| Coulombic effects in the solid phase | Ionic solids are held together by electrostatic forces. For regular crystals, the molar potential energy of an ion is given by |
|---|---|

$$V = A \frac{N_A e^2}{4\pi\varepsilon_o} \left( \frac{z_A z_B}{d} \right)$$

The Madelung constant, A, depends only upon the symmetry of the ions in the structure, and hence on the structure type. The scaling distance $d$ varies in proportion to the size of the unit cell.

**Related topics**  The wave nature of matter (G4)  Molecular orbital theory of
  Statistical thermodynamics (G8)      diatomic molecules II (H4)
  Molecular orbital theory of
    diatomic molecules I (H3)

**Bonding in solids: band theory**

Bonding in solids involves orbital contributions from far more atoms than are encountered in most molecular systems. Far from complicating the bonding theory, this very large number enables the bonding to be treated by averaging of all the possible bonding patterns. Experiments demonstrate that the bonding in solids does not yield discrete energy levels, but leads to the formation of energy **bands** within which a given electron may hold any energy within a continuous range. The model of these bands is referred to as **band theory.** Two complementary theories form the basis of band theory. The **tightly bound electron model** (also known as the **tight binding approximation**) and the **free electron model**.

**Tightly bound electron model**

The tightly bound electron model is an extension of the LCAO approach to molecular orbitals. The energy bands in a solid are treated as a linear combination of the atomic orbitals of its consituent elements.

The standard approach analyses the bonding in the hypothetical case of a linear chain of hydrogen atoms (*Fig. 1a*). As the number of atoms in the chain are increased the number of molecular orbitals increases also. If the chain consists of N hydrogen atoms, the lowest energy molecular orbital is formed when all the hydrogen 1s atomic orbitals are in phase (i.e. the wavefunctions all have the same sign). The highest energy orbital is formed when all the atomic orbitals are out of phase (i.e. the sign of the atomic wavefunctions alternates between + and –). In between are N-2 molecular orbitals, whose energies depend upon the phase of the component orbitals. If the value of N is very large, the molecular orbitals effectively form a continuous **band**. With large numbers of atoms the energy gap between orbitals is only of the order of $10^{-40}$ J, effectively making the band into an energy continuum. The energy difference between the upper and lower orbitals is known as the **band width,** and increases progressively more slowly as more orbitals are added.

The number of molecular states per unit binding energy is referred to as the **density of states**. *Figure 1a* illustrates that the density of the states is higher towards the edges of the band than in the middle, and the corresponding density of states diagram is of the form shown in *Fig. 1b*.

In real solids the bands are formed in three dimensions using p, d, and f orbitals in addition to the s orbitals, and both the structure of the bands and the density of states diagrams are far more complex than in the case of a one-dimensional material.

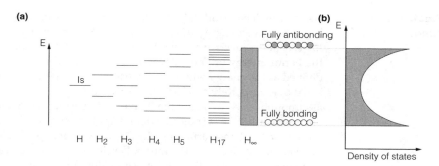

Fig. 1.   (a) Formation of a band from a chain of hydrogen atoms. (b) The resulting density of states.

**Free electron model**

The free electron model makes no assumptions about the molecular state of matter, but assumes that the electrons in a metal are free to move throughout the available volume unhindered. This approach reduces the problem of the energy levels to that of a particle in a box (see Topic G4). This treatment shows that the number of electrons, $N$, of mass, $m_e$, which may be accommodated in energy levels up to a maximum energy of $E_{max}$ within a three-dimensional cube of sides, $a$, is given by:

$$N = \frac{8\pi a^3}{3(2m_e E_{max}/h^2)^{3/2}}$$

This gives a **density of states diagram** which has the form shown in *Fig. 2*. The equation may also be modified to calculate the maximum energy of the electrons from their number density, in the solid. As $a^3$ is the volume of the box, the number density of the electrons, $\rho$, is equal to $N/a^3$, and $E_{max}$ is given by:

$$E_{max} = \frac{h^2}{2m_e(3\rho/8\pi)^{2/3}}$$

The free electron model is remarkably accurate in calculations of electron energies in simple metals. The major difference between the free electron model and the tightly bound electron model is that in the free electron model the band is open-ended, and does not have an upper energy limit. More accurate approaches to band theory account for both the highly delocalized nature of the electrons, and the regular array of nuclear potentials.

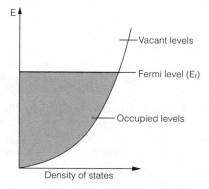

Fig. 2.   Density of states diagram resulting from the free electron model.

**Electrons in bands**

As with molecular orbitals, the energy levels in a band are progressively filled from the lowest energy upwards. At zero Kelvin, the electrons occupy only the lowest energy levels available. The highest occupied energy level is referred to as the **Fermi energy** or **Fermi level**, $E_f$ (*Fig. 2*). The Fermi energy is more precisely defined as the energy where the probability of it being occupied by an electron is ½. Above zero Kelvin, electrons are thermally excited to higher levels. The assumptions used to obtain the Boltzmann distribution for atomic and molecular energy distributions (Topic G8) do not apply to electrons in solids. With the limitations of the band structure, the Pauli exclusion principle, and the indistinguishability of electrons, the thermal excitation is best described by **Fermi-Dirac statistics.** The probability $f(E)$ of occupation for an electron at an energy $E$ at a temperature, $T$ is given by:

$$f(E) = \frac{1}{1 + \exp\left[\left(E - E_f\right)/k_B T\right]}$$

The resulting form of the electron distribution is illustrated for an idealized band in *Fig. 3*.

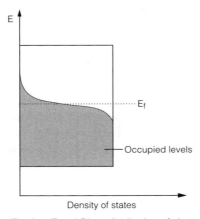

*Fig. 3.    Fermi-Dirac distribution of electrons in an idealized band.*

**Metals, insulators and semiconductors**

In real solids, a number of different bands are formed from the interaction of a number of atomic orbitals. The relative energy of these bands and the number of electrons which occupy them dictate the electrical properties of the solid.

The conductivity of a solid is proportional to the number and mobility of the charge carriers. In a **metal**, or a **metallic conductor**, one or more of the bands is only partially filled, giving the electrons vacant orbitals into which they may move freely. In the simplest cases, this results from an insufficient number of electrons being available from the constituent atoms to fill the band. In the metal sodium, for example, $N$ atoms each donate one electron into a 3$s$ band which requires 2$N$ electrons to be fully occupied (*Fig. 4a*). Metallic behavior may also result when a band is prevented from being filled by the overlap of a second band. This occurs in, for example, magnesium, where $N$ atoms each donate 2$N$ electrons into the 3$s$ band. The 3$s$ band is not filled, as it is overlapped by the 3$p$ band. Electrons enter the lower levels of the $p$ band instead of completely filling the $s$ band, leading to metallic conduction (*Fig. 4b*).

In an **insulator**, the highest occupied band (the **valence band**) is filled, and is energetically separated from the lowest unoccupied band (the **conduction**

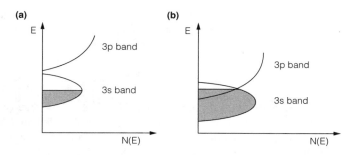

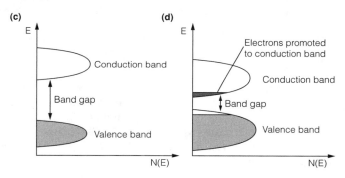

Fig. 4. Idealized band structures. (a) Metallic conductor with partially filled band resulting from partially filled atomic orbitals; (b) metallic conductor with partially filled band resulting from band overlap; (c) an insulator; (d) a semiconductor.

band) by a **band gap** (*Fig. 4c*). The band gap is large enough to prevent thermal excitation of electrons into the conduction band. Electrical conduction is prevented because the electrons do not have vacant energy levels into which they may migrate, and so cannot carry charge through the material.

A **semiconductor** has a similar overall band structure to that of an insulator, and there is no clear distinction between the two. What difference there is lies in the size of the band gap, which is small enough in a semiconductor to allow thermal excitation of electrons into the conduction band (*Fig. 4d*). For materials which are typically regarded as semiconductors at room temperature (silicon, gallium arsenide, gallium nitride, for example), the band gap is of the order of 1–3 eV. Raising the temperature of a semiconductor increases the number of electrons promoted across the band gap.

The conductivity of a semiconductor increases with temperature as more electrons are promoted into the conduction band; this behavior contrasts with the behavior of a metal, whose conductivity decreases with temperature. Metallic behavior is explained by the increased thermal motion of the lattice, which limits the electron mobility by increasing the electron–lattice collision frequency.

**Coulombic effects in the solid phase**

In ionic solids, ions are primarily held together by electrostatic forces. The strength of the binding in these solids is measured in terms of the energy required to fully dissociate the ions in the lattice into gaseous ions. The energy required for this process to be brought about is the **lattice enthalpy** (see Topic B3).

If $e$ is the charge on an electron, and the number of charges on a pair of ions, $A$ and $B$, held a distance $r_{AB}$ apart are $z_A$ and $z_B$ respectively, then the electrostatic bonding energy between the pair may be calculated:

$$V = \frac{z_A z_B e^2}{4\pi\varepsilon_o r_{AB}}$$

where $\varepsilon_o$ is the vacuum permittivity. The electrostatic energy of an ion in an infinite one-dimensional array of ions may be easily calculated by summation of the individual terms, but calculation of the potential energy in higher-dimensional lattices is mathematically challenging. In order to calculate the total coulombic potential for an ion, the contribution from every other ion is required. The importance of the strength of each interaction decreases with distance, but the summation is made difficult because the number of interactions increases with distance. It is found that, for regular crystals, the molar potential energy is given by:

$$V = A\frac{N_A e^2}{4\pi\varepsilon_o}\left(\frac{z_A z_B}{d}\right)$$

The energy depends upon the charges on the ions (larger charges giving higher energy), a scaling distance, $d$, and the **Madelung constant, A**. The Madelung constant depends only upon the symmetry of the ions in the structure, and hence on the structure type. Thus A = 1.763 for all cesium chloride lattices and A = 2.519 for all fluorite lattices. The scaling distance $d$ varies in proportion to the size of the unit cell.

# H6 WEAK INTERMOLECULAR INTERACTIONS

## Key Notes

**Weak intermolecular forces**

The forces between molecules typically involve energies of significantly less than 50 kJ mol⁻¹, and are regarded as weak.

**Dipolar interactions**

When a material consists of molecules with permanent dipoles, and the thermal energy is sufficient to ensure random orientation of the dipoles, the potential experienced by the interaction of two dipoles is proportional to $r^{-6}$, where $r$ is the interdipole distance.

**Polarizability**

The dipole moment induced in a polarizable molecule through the effect of an electric field is referred to as an induced dipole moment. The strength of the induced dipole is equal to the product of the polarizability, $\alpha$, and the electric field strength, $E$. Where two molecules have a permanent dipole, the energy of the interaction is proportional to $r^{-6}$.

**London dispersion interaction**

A spontaneous electric dipole in one molecule may induce an electric dipole moment in a second molecule, to give an attractive potential known as the dispersion interaction. Dispersion forces are always attractive, and are independent of temperature.

**Repulsion energy**

At small intermolecular distances, the electron clouds of the molecules begin to interpenetrate, and a very strong repulsion energy becomes important, best written in the form $U(r) = +\beta\exp(-r/\rho)$, where $\beta$ and $\rho$ are empirical factors. The repulsive force is insignificant at high $r$, yet dominant at low $r$.

**The total intermolecular interaction energy**

The total interaction between two molecules may be fitted to the expression:

$$U(r) = \frac{b}{r^{12}} - \frac{C}{r^6}$$

The resulting curve is known as the Lennard-Jones potential. The collision diameter, $\sigma$, corresponds to the intermolecular distance at which $U(r) = 0$.

**Hydrogen bonding**

Hydrogen bonds are a very specific form of intermolecular bonding formed when a hydrogen atom is covalently bonded to the strongly electronegative elements oxygen, nitrogen or fluorine. In molecular orbital terms, the hydrogen decreases the energy of the bonding orbitals between the two electronegative elements, by the inclusion of an extra term resulting from the hydrogen 1s orbital, in the linear combination of molecular orbitals.

**Related topic**

Strong solid phase interactions (H5)

**Weak intermolecular forces**

The distinction between **weak** and **strong binding energies** is an arbitrary one, but binding energies of >200 kJ mol$^{-1}$, limited to covalent or ionic bonds, are generally regarded as strong. The forces between molecules are regarded as weak with typical energies significantly less than 50 kJ mol$^{-1}$.

**Dipolar interactions**

When a material consists of molecules with **permanent electric dipoles** (see Topic H5), the electrostatic force between molecules changes their potential energy. The energy, $U(r)$, of the electrostatic interaction between two parallel dipoles with dipole moments $\mu_1$ and $\mu_2$, a distance $r$ apart, is given by:

$$U(r) = \frac{\mu_1\mu_2}{2\pi\varepsilon_o r^3}\left(1 - \cos^2\theta\right)$$

where $\theta$ is the angle between the dipole axes and the vector linking the centers of the two dipoles. This potential is a maximum if $\theta = 0°$, when the positive pole of one dipole experiences only the negative pole of the second. The potential is zero if $\theta = 90°$, as both negative and positive poles of one dipole are experienced equally, and so cancel out, at the second dipole.

In real systems, the strength of the interaction is usually less than, or comparable with, the thermal energy, and the dipoles are able to rotate in three dimensions. At low temperatures, the thermal energy is insufficient to overcome some alignment of the dipoles, whereas higher temperatures have the effect of averaging out the potentials of the surrounding dipoles. Under these conditions, the expression for the potential experienced by a dipole, at temperature $T$, alters to:

$$U(r) \propto -\frac{\left(\mu_1\mu_2\right)^2}{Tr^6}$$

**Polarizability**

Regardless of whether or not a molecule or atom possesses a permanent dipole moment, a dipole moment, resulting from charge separation, may be induced in the molecule through the effect of an electric field. Such a molecule is **polarizable** and the resulting dipole is referred to as an **induced dipole moment.** The strength of the induced dipole is proportional to the strength of the electric field, $E$:

$$\mu = \alpha E \qquad \text{where } \alpha = \frac{\varepsilon - 1}{4\pi N'}$$

$\alpha$ is referred to as the **polarizability**, $\varepsilon$ is the dielectric constant, and $N'$ the number of dipole moments per cm$^3$. Where the induced dipole moment arises as a result of a molecule, of polarizability $\alpha_1$, with a permanent electric dipole, $\mu_2$, an attractive potential is set up between the two molecules whose strength is given by:

$$U(r) \propto -\frac{\alpha_1\mu_2^2}{r^6}$$

Where both molecules have a permanent dipole, the energy of the interaction is given by:

$$U(r) \propto -\frac{1}{r^6}\left(\alpha_1\mu_2^2 - \alpha_2\mu_1^2\right)$$

**London dispersion interaction**

Electrostatic interactions may occur between atoms and molecules which do not contain polar bonds. Such forces are responsible for the existence of solid and liquid helium and methane, for example, and are the result of the polarizability of the molecule. At any one instant, the electron density around a molecule need not be spherically distributed, and this asymmetry has a dipole associated with it. This spontaneous electrical dipole may in turn induce an electric dipole moment in a second molecule, to give an attractive potential (*Fig. 1*). This phenomenon is referred to as the **London interaction** or the **dispersion interaction**.

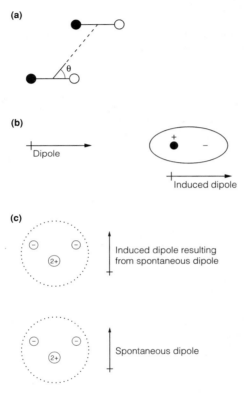

*Fig. 1. Schematic representation of (a) the dipolar interaction between parallel permanent dipoles, (b) a permanent and an induced dipole, and (c) the dispersion interaction between two helium atoms.*

The dispersion interaction energy of a pair of isotropic molecules with ionization energies $hv_1$ and $hv_2$, separated by a distance $r$ is given by:

$$U(r) \propto -\frac{3}{2}\frac{h}{r^6}\left(\frac{v_1 v_2}{v_1 + v_2}\right)\alpha_1\alpha_2$$

These dispersion forces are always attractive, and are independent of temperature. At sufficiently low temperatures, therefore, all molecular and atomic substances must condense to either the liquid or solid phase when $kT$ is less than the dispersion energy.

**Repulsion energy**

At small intermolecular distances, the electron clouds of the molecules begin to interpenetrate, and a very strong repulsion energy becomes important. It is possible to express this repulsion term in two forms:

$$U(r) = +\frac{b}{r^n} \quad \text{or} \quad U(r) = +\beta \exp(-r/\rho)$$

where $b$, $\beta$, $n$ and $\rho$ are empirical factors whose value is chosen to best fit the data. The exponential expression is both theoretically and experimentally preferred, but the differences are minimal. In the former expression, the radius is generally raised to its twelfth power (i.e. $n = 12$), making the repulsive force insignificant at high $r$, yet dominant at low $r$.

**The total intermolecular interaction energy**

For the case of dipolar, polarizable, molecules in the gas phase, the total intermolecular interaction energy is approximately given by the sum of the attractive and repulsive interaction energies:

$$U(r) = \underset{\text{repulsion}}{\frac{A}{r^{12}}} - \underset{\text{dipole-dipole}}{\frac{(\mu_1\mu_2)^2}{Tr^6}} - \underset{\text{induced dipole}}{\frac{\alpha_1\mu_2^2}{r^6}} - \underset{\text{dispersion}}{\frac{3}{2}\frac{h}{r^6}\left(\frac{v_1 v_2}{v_1 + v_2}\right)\alpha_1\alpha_2}$$

where A is a constant. The total interaction simplifies considerably to the expression:

$$U(r) = \frac{b}{r^{12}} - \frac{C}{r^6}$$

where the constant, $C$, is the sum of the coefficients in the expressions for the attractive potentials. The resulting curve is known as the **Lennard-Jones potential** (*Fig. 2*). This plot also enables definition of a **collision diameter**, $\sigma$, which corresponds to the intermolecular distance at which $U(r) = 0$.

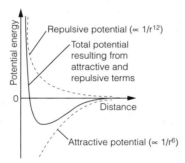

*Fig. 2. The Lennard-Jones potential.*

**Hydrogen bonding**

**Hydrogen bonding** is a very specific form of intermolecular bonding which gives rise to an attractive intermolecular potential, typically with a value of some tens of kJ mol⁻¹. Hydrogen bonds are formed when a hydrogen atom is covalently bonded to the strongly electronegative elements oxygen, nitrogen or fluorine. The effect is that of a strong attractive force between the hydrogen on one molecule, and the oxygen, nitrogen, or fluorine, on another. The effect is often inadequately described in terms of electrostatic attraction.

In molecular orbital terms, the hydrogen decreases the energy of the bonding orbitals between the two electronegative elements, by the inclusion of an extra term, from the hydrogen 1s orbital, in the linear combination of molecular orbitals. In the absence of a hydrogen atom, a linear combination of two fluorine orbitals forms one bonding and one antibonding orbital. The four available electrons fill these orbitals to give no net bonding. Inclusion of the hydrogen $1s$ orbital into the linear combination creates a third non-bonding combination, to give a bonding interaction overall (*Fig. 3*).

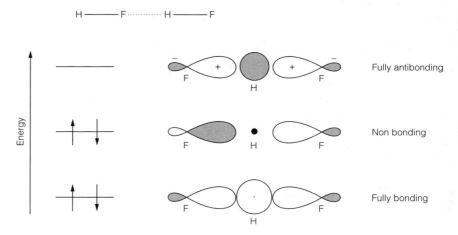

Fig. 3.    Energy levels in a hydrogen bond ($\cdots\cdots\cdots$).

# I1 GENERAL FEATURES OF SPECTROSCOPY

## Key Notes

**Spectroscopy**
Spectroscopy is the analysis of the electromagnetic radiation emitted, absorbed or scattered by atoms or molecules as they undergo transitions between two discrete energy states.

**The electromagnetic spectrum**
Electromagnetic radiation consists of a propagating oscillating electric and magnetic field. The frequency of oscillation defines different regions of the electromagnetic spectrum: radio waves, microwaves, infrared, visible, ultraviolet, X-rays and γ-rays. The energy of electromagnetic radiation of frequency, $v$, is quantized in units of $hv$ called photons.

**Selection rules**
A selection rule states whether a particular spectroscopic transition is allowed or forbidden. A physical selection rule describes the general properties the molecule must possess in order to undergo a certain class of transitions. A specific selection rule states what changes in quantum number are allowed for a transition to occur.

**Emission, absorption and scattering spectroscopy**
Emission spectroscopy is the analysis of the energy of photons emitted when a molecule moves from a higher to a lower energy state. Absorption spectroscopy is the analysis of the energy of photons absorbed from incident light by a molecule undergoing transition from a lower to a higher energy state. Scattering spectroscopy is the analysis of the energy lost or gained by a photon of incident light after it has undergone an energy exchange interaction with a molecule.

**Related topics**
Chemical and structural effects of quantization (G7)

Practical aspects of spectroscopy (I2)

---

**Spectroscopy**

**Quantum theory** shows that atoms and molecules exist only in discrete states, each of which possess discrete values, or **quanta**, of energy. The states are called the **energy levels** of the atom or molecule. **Spectroscopy** is the analysis of the electromagnetic radiation emitted, absorbed or scattered by atoms or molecules as they undergo **transitions** between two energy levels. The frequency, $v$, of the **electromagnetic radiation** associated with a transition between a pair of energy levels $E_1$ and $E_2$ is given by:

$$hv = |E_1 - E_2| = |\Delta E|$$

Spectroscopy that probes different magnitudes of energy level separation is associated with different regions of the **electromagnetic spectrum**.

The analysis of **atomic spectra** (see Topics G5 and G7) yields information about the electronic structure of the atom. Molecules also possess energy due to rotation and to the vibration of the bonds between the atoms, so **molecular**

**spectra** are more complicated, since they include rotational and vibrational transitions as well as electronic transitions. However, the analysis of molecular spectra provides a wealth of information on molecular energy levels, bond lengths, bond angles and bond strengths. The characteristic spectral frequencies associated with particular atoms and molecules also means that spectroscopy is widely used for identification of species and monitoring of specific reactants or products in, for example, kinetic measurements (see Topic F1).

## The electromagnetic spectrum

**Electromagnetic radiation** is a propagating oscillation of interconnected electric and magnetic fields. The fields oscillate in phase along the direction of propagation as sine waves with frequency, $v$, and wavelength, $\lambda$, related by $c = v\lambda$, where $c$ is the speed of light in a vacuum ($c = 2.9979246 \times 10^8$ m s$^{-1}$). All electromagnetic waves travel at this speed. The frequency of electromagnetic radiation is often specified as **wavenumber**, $\tilde{v}$:

$$\tilde{v} = \frac{1}{\lambda} = \frac{v}{c}$$

with units of reciprocal centimeters, cm$^{-1}$. Individual **photons** of electromagnetic radiation have energy, $E$:

$$E = hv = hc\tilde{v} = \frac{hc}{\lambda}$$

i.e. photon energy is proportional to frequency and inversely proportional to wavelength. The wavelength range of electromagnetic radiation encompasses many orders of magnitude, from $\sim10^3$ m at the low frequency, low energy end, to $\sim10^{-12}$ m at the high frequency, high energy end (*Fig. 1*). Different regions of the spectrum correspond to different types of radiation. For example, radiation visible to the human eye occurs over a very narrow range of wavelengths between about 700 and 400 nm.

The interaction of the oscillating electric and magnetic fields with the electrical and magnetic properties of atoms and molecules gives rise to various forms of spectroscopy. Because of the direct relationship between energy and radiation frequency (or wavelength), spectral transitions between different types of atomic or molecular energy levels are associated with different regions of the **electromagnetic spectrum**. For example, the rotational energy levels of molecules are more closely spaced than the vibrational energy levels which are more closely spaced than the electronic energy levels. Therefore, **rotational**,

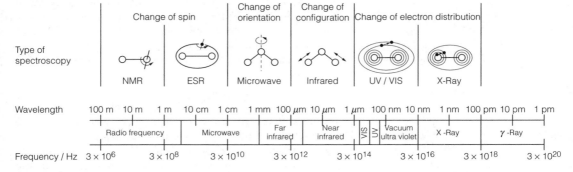

*Fig. 1.   The electromagnetic spectrum and the type of spectroscopy associated with different spectral regions.*

**vibrational** and **electronic spectroscopy** are usually associated with electromagnetic radiation in the microwave, infrared and visible/ultraviolet regions of the spectrum, respectively (*Fig. 1*).

**Selection rules**

Whether a **transition** between a particular pair of energy levels is **allowed** or **forbidden**, is determined by the **selection rule(s)** for the type of atomic or molecular energy under consideration.

A **physical selection rule** specifies the general features that the molecule must have in order to exhibit any non-zero **transition probabilities** for a particular type of energy transition. The magnitude of the transition probability determines the **intensity** of a transition (see Topic I2).

A **specific selection rule** specifies exactly which pairs of quantum states are linked by allowed transitions, provided that transitions are allowed by the physical selection rule.

**(a)**

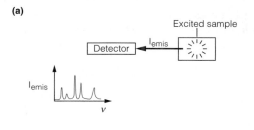

**(b)**

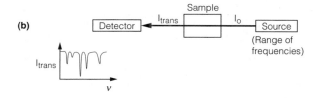

**(c)**

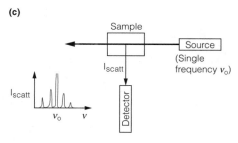

*Fig. 2. Schematic representation of (a) emission, (b) absorption and (c) Raman spectroscopy.*

**Emission, absorption and scattering spectroscopy**

The different types of spectroscopy are illustrated schematically in *Fig. 2*. In **emission spectroscopy**, a molecule (or atom) already in an excited state undergoes a transition from a state of high energy to a state of low energy, and emits the excess energy as a photon. The distribution in frequency of the emitted photons is the emission spectrum.

In **absorption spectroscopy**, a photon of specific frequency is absorbed by the molecule to promote it from a low energy state to a high energy state. The absorption spectrum is obtained by observing the intensity of the transmitted radiation relative to the incident radiation over a range of frequencies of incident light.

In the **Raman** variant of **scattering spectroscopy**, a monochromatic (single frequency) beam is directed at the sample and the frequency of the light that is scattered away from the direction of the incident beam is analyzed. A proportion of the scattered photons have different frequency to the incident photons because the molecule gains or loses energy in the collision. Photons that lose energy in the interaction travel away at lower frequency to the incident light (**Stokes scattering**); photons that acquire energy from the interaction travel away at higher frequency (**anti-Stokes scattering**).

Emission, absorption and Raman spectroscopy all provide essentially the same information about energy level separations, but practical considerations and **selection rules** generally dictate which technique is most appropriate. Absorption spectroscopy is usually the most straightforward to apply.

# 12 PRACTICAL ASPECTS OF SPECTROSCOPY

## Key Notes

**Experimental apparatus**

All spectroscopic measurements require a radiation source (for emission spectroscopy the excited sample acts as its own source), a dispersing element (to separate radiation into its component frequencies), and a detector (to measure radiation intensity). The exact nature of these components depends on the region of the electromagnetic spectrum under study.

**Intensity of spectral lines**

The intensity of a spectral transition is proportional to the transition probability, the concentration of molecules in the initial state of the transition, and (for absorption measurements) the path length of the radiation through the sample. The transition probability is a property intrinsic to the particular pair of initial and final states.

**Beer-Lambert law**

The Beer-Lambert law, $\log (I/I_0) = -\varepsilon[X]l$, describes the exponential decrease in the transmittance, $I/I_0$, of light through an absorbing sample, where $I$ is the intensity of transmitted light, $I_0$ is the intensity of incident light, $l$ is the path length, $[X]$ is the sample concentration, and $\varepsilon$ is the absorption coefficient. The quantity $-\log (I/I_0)$ is called the absorbance.

**Linewidths**

A spectral transition is never infinitely narrow because of the uncertainty in energy that is intrinsic to the finite lifetime of all excited states. The shorter the lifetime, the greater the energy uncertainty in the spectral line. This natural linewidth is often exceeded by the collisional linewidth which arises when the lifetime of the excited state is decreased by molecular collisions that remove energy non-radiatively. The Doppler effect also contributes to linewidth for gaseous samples.

**Lasers**

Laser (light amplification by the stimulated emission of radiation) action occurs when a radiative transition is stimulated from an upper state that has greater population than the lower state (a population inversion). Laser radiation is intense, monochromatic and unidirectional.

**Related topic**

General features of spectroscopy (I1)

---

**Experimental apparatus**

*Figure 2* in Topic I1 illustrates the basic experimental arrangements for **emission**, **absorption** and **Raman** spectroscopic measurements. The three basic components of a spectroscopic apparatus are: a **radiation source** (not required for emission spectroscopy), a **frequency dispersing element**, and a **detector**.

*Radiation source*

The **radiation source** for absorption spectroscopy must emit over a range of frequencies and depends on the region of the **electromagnetic spectrum** under investigation. For the ultraviolet region, a discharge through deuterium or xenon gas in a quartz cell is used as a broad-band source; for the visible region a tungsten-iodine lamp is used. Far infrared radiation is provided by a mercury discharge inside a quartz tube, and near infrared radiation is usually provided by a heated ceramic filament. Microwave radiation is generated by a device called a **klystron**. Radio-frequency radiation (as required for **nuclear magnetic resonance spectroscopy**) is generated by oscillating an electric current through coils of wire at the appropriate frequency.

The intense monochromatic light source required for Raman spectroscopy is usually provided by a visible or ultraviolet **laser**.

*Dispersing element*

The **dispersing element** separates the emitted, transmitted, or scattered radiation into its constituent frequencies after interaction with the sample under investigation. (In some applications of **absorption spectroscopy**, the dispersing element is used to separate the broad band source into frequency-resolved radiation before it is incident on the sample.) The simplest dispersing element is a glass or quartz prism. Diffraction gratings are also widely used. These consist of parallel lines etched into the surface of a glass or ceramic plate at spacings comparable with the wavelength of the radiation being dispersed. Radiation incident onto the surface of the grating is reflected (dispersed) at different angles according to incident frequency because of destructive and constructive wave interference.

*Detector*

The **detector** is a device that produces an electrical voltage or current in response to the intensity of incident radiation. In the visible and ultraviolet regions photomultiplier tubes are widely used. An incident photon ejects an electron from a photosensitive surface, the electron is accelerated by a potential difference to strike another surface, and the shower of secondary electrons from this collision is accelerated towards another surface, and so on. Thus, each photon creates an amplification electron cascade which is converted into an electric current. Infrared detectors often consist of a mixture of metal alloys or a mixture of solid oxides whose electrical resistances change as a function of temperature. Throughout the visible and infrared regions of the spectrum, specific radiation-sensitive semiconductor devices have been developed which convert incident photons directly into an electrical signal.

*Sample presentation*

In **absorption spectroscopy**, the extent of absorption depends, amongst other factors, on the **path length** of the incident radiation through the sample. The path length required for gaseous samples is often longer than for liquid samples because the concentration is usually lower. Gaseous samples are essential for rotational (microwave) spectroscopy in order that the molecules can rotate freely. For vibrational (infrared) spectroscopy, liquid or solid samples are often ground to a paste with 'Nujol', a hydrocarbon oil, and held between sodium chloride or potassium bromide windows which are transparent to frequencies of 700 cm$^{-1}$ and 400 cm$^{-1}$, respectively. The path length in any form of absorption spectroscopy can be increased by reflecting the incident beam multiple times through the sample using mirrors at each end of the sample cavity.

**Intensity of spectral lines**

Three principal factors influence the **intensity** of a spectral transition:

(i) The **transition probability**. This property is determined by the nature of the initial and final quantum states of the molecule. Although detailed calculation of absolute transition probabilities is often complex, it is usually possible to derive general **selection rules** (see Topic I1) that distinguish whether a transition probability is zero (**forbidden**) or non-zero (**allowed**). The magnitude of a transition probability may be determined experimentally from absorption spectroscopy by application of the **Beer-Lambert law**.

(ii) The concentration of the initial state. The greater the concentration of molecules in the initial energy level of a transition, the more intense the spectral transition.

(iii) The path length of the sample. For absorption transitions, the more sample the beam of radiation traverses, the more energy will be absorbed from it.

**Beer-Lambert law**

The **Beer-Lambert law** states that the intensity of radiation absorbed by a sample is proportional to the intensity of the incident radiation, $I_0$, the concentration of the absorbing species, $[X]$, and the path length of the radiation through the sample, $l$. Mathematically, the observation can be written in terms of the decrease in intensity, $-dI$, that occurs for an increase in path length $dx$:

$$-dI = \sigma I[X]dx$$

where $\sigma$ is the constant of proportionality, which depends on the identity of the absorbing species and the frequency of the incident radiation. Rearranging the expression and integrating over the full path length (from 0 to $l$) along which absorbance occurs gives:

$$\int_{I_0}^{I} \frac{dI}{I} = -\sigma[X] \int_0^l dx$$

$$\ln \frac{I}{I_0} = -\sigma[X]l$$

i.e. the Beer-Lambert law for the intensity, $I$, of transmitted radiation is:

$$I = I_0 \exp(-\sigma[X]l)$$

The constant $\sigma$ is called the **absorption coefficient** or **extinction coefficient**.

The Beer-Lambert law is often written in terms of logarithms to the base 10, in which case, by writing $\sigma = \varepsilon \ln 10$:

$$\log \frac{I}{I_0} = -\varepsilon[X]l$$

The constant $\varepsilon$ is another form of the absorption (extinction) coefficient. The constants $\sigma$ and $\varepsilon$ are related directly to the **transition probability** for the spectral transition. The dimensions of $\sigma$ and $\varepsilon$ are (concentration × length)$^{-1}$ but the exact units depend on the units used for the species concentration (e.g. molar or molecular units) and path length. Care must be taken to determine whether values of absorption coefficient are referenced to logarithms of absorption intensity ratios to base $e$ or to base 10. The convention is as written here. In either form, it can be seen that transmitted intensity decreases exponentially with the length of sample through which the radiation passes.

The ratio of the transmitted intensity to the incident intensity, $I/I_0$, is called the **transmittance,** $T$, so log

$$\log T = -\varepsilon[X]l$$

The **absorbance** (or **optical density**) of the species,

$$A = -\log\frac{I}{I_0}$$

is related to transmittance through, $T = 10^{-A}$.

**Linewidths**

Spectral lines are not infinitely narrow since that would violate a variant of the **Heisenberg uncertainty principle** (Topic G4) which states that the energy of a state existing for a time, $\tau$, is subject to an uncertainty, $\delta E$, of magnitude:

$$\delta E \approx \frac{\hbar}{\tau}$$

Since no excited state has an infinite lifetime, the spectral transition corresponding to the energy separation between two states is spread over a finite width of energy. The energy uncertainty inherent to states that have finite lifetimes is called **lifetime broadening**. Two processes contribute to the finite lifetime of excited states:

(i)   The rate of spontaneous emission of radiation as an excited state collapses to a lower state (a fundamental property of the molecule) establishes an intrinsic minimum **natural linewidth** to the transition, $\delta E_{nat} \approx \hbar/\tau_{nat}$ where $\tau_{nat}$ is the natural lifetime to spontaneous decay.

(ii)  The rate of collisions of the molecules with each other and the walls of the container establishes a **collisional linewidth**, $\delta E_{coll} \approx \hbar/\tau_{coll}$ where $\tau_{coll}$ is the average time between the deactivation collisions. For liquids, and gases at moderate pressures, the collisional linewidth dominates the natural linewidth.

A third line-broadening process, particularly important for gaseous samples, is the **Doppler effect**, in which radiation shifts to higher (or lower) frequency when the source is moving towards (or away) from the observer, respectively. Since molecules in a sample are moving in all directions with respect to the detector, with a range of velocities given by the **Maxwell distribution** (Topic A2), each spectral transition is spread over a range of Doppler frequency shifts. **Doppler linewidth** increases with temperature because the molecules have a greater range of speeds.

**Lasers**

The word **laser** is an acronym for light amplification by the stimulated emission of radiation. Laser action requires: (i) the production of a **population inversion**, in which the population in an upper (excited) state exceeds the population in a lower state, and (ii) the stimulation of a radiative transition between the two states. The excited state is stimulated to emit a photon by interaction with radiation of the same frequency. The more photons of that frequency present, the greater the number of photons the excited states are stimulated to emit. This positive feedback process is known as the **gain** of the laser medium.

A greater population is required in the upper state in order to ensure that net emission rather than net absorption occurs. The population inversion must be prepared deliberately (a process called **pumping**) because the **Boltzmann**

**distribution law** dictates that population is greater in the lower energy state at thermal equilibrium. One way of achieving this is to pump an excited level (using an intense flash of light from a discharge or another laser) which converts non-radiatively into the upper level of the laser transition (*Fig. 1*). In the four-level system illustrated the laser transition terminates in another excited level and population inversion is easier to achieve than when the laser transition terminates in the ground state. Continuous rather than pulsed laser output is possible if the population inversion can be sustained.

The characteristics of laser radiation are that it is: (i) intense, (ii) **monochromatic** (narrow frequency), (iii) **collimated** (low beam divergence), and (iv) **coherent** (all waves in phase).

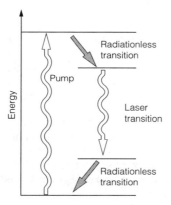

*Fig. 1.    The transitions involved in a four-level system operating as a laser.*

# I3 ROTATIONAL SPECTROSCOPY

## Key Notes

| | |
|---|---|
| **Rotational energy levels** | The allowed energy levels of a rigid linear rotor are $E_J = BJ(J + 1)$ where $J$ is the rotational quantum number, and $B$ is the rotational constant, which is inversely proportional to the moment of inertia. The energy levels of a symmetric top molecule (one axis of rotational symmetry) are described by two quantum numbers, $J$ and $K$, and two rotational constants $B$ and $A$. |
| **Microwave rotational spectroscopy: selection rules and transitions** | A molecule only gives rise to a rotational spectrum if it possesses a permanent electric dipole. The specific selection rule allows transitions of $\Delta J = \pm 1$ and $\Delta K = 0$. The allowed transitions in the rotational spectrum of a polar linear and symmetric top molecule have energy $2B(J + 1)$ and generally fall within the microwave region of the electromagnetic spectrum. |
| **Rotational intensities** | A rotational spectrum has a characteristic intensity distribution that passes through a maximum because the population of rotational levels from which the spectral transitions originate is proportional to the Boltzmann factor (declines with rotational quantum number, $J$) and the degeneracy of the rotational level (increases with $J$). The rotational level with maximum population is $$\sqrt{\frac{k_B T}{2B}} - \frac{1}{2}.$$ |
| **Rotational Raman spectroscopy: selection rules and transitions** | The polarizability of the molecule must be anisotropic to give rise to a rotational Raman spectrum. The specific selection rule for linear molecules allows transitions of $\Delta J = \pm 2$. The anti-Stokes and Stokes rotational transitions occur at energies $\pm 2B(2J + 3)$ from the energy of the incident excitation radiation. |
| **Related topics** | The wave nature of matter (G4)    Practical aspects of spectroscopy<br>General features of spectroscopy (I1)    (I2) |

**Rotational energy levels**

Application of **quantum theory** shows that the rotational energy possessed by a molecule is **quantized**, in the same way that all energy is quantized. In general, the rate of rotation of a molecule is sufficiently slow compared with the rate of vibration of the bonds, that the molecule can be considered as a **rigid body** rotating with fixed internuclear separations given by the average of the vibrational displacements.

*Diatomic molecule*
The simplest type of rigid rotor is the linear rotor of a diatomic molecule. When the **Schrödinger equation** is solved for a linear rigid rotor (see **particle in a circular orbit**, Topic G4), the allowed energy levels turn out to be quantized according to:

$$E_J = BJ(J + 1) \qquad J = 0,1,2,...$$

where $J$ is the **rotational quantum number**, and

$$B = \frac{\hbar^2}{2I}$$

is called the **rotational constant**. The distribution of energy levels is illustrated in *Fig. 1*. The separation between energy level increases with energy. Since $J$ can take the value zero, molecules have no rotational **zero point energy**. The parameter $I$ is the **moment of inertia** of the molecule, and for a diatomic AB of equilibrium internuclear separation $R_e$ is given by, $I = \mu R_e^2$ where $\mu = m_A m_B / (m_A + m_B)$ is the **reduced mass** of the bond.

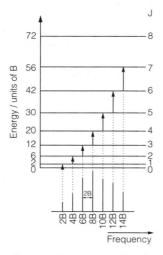

Fig. 1.    *The energy levels of a linear rigid rotor showing the allowed rotational transitions of a polar linear molecule.*

The moment of inertia is the rotational equivalent of linear momentum. A body with a large moment of inertia requires a greater twisting force to reach a certain rate of rotation and possesses greater energy for a given rate of rotation than a body with a small moment of inertia. Since $B$ is inversely proportional to $I$, the larger the moment of inertia (corresponding to a longer bond and heavier atoms) the smaller the rotational constant and the more closely spaced the rotational energy levels.

The energies of rotational levels are usually quoted in units of reciprocal centimeter, or **wavenumber**, cm⁻¹, instead of joules. The rotational constant in units of wavenumber is:

$$B = \frac{\hbar}{4\pi I c}$$

*Polyatomic molecules*

The rotational energy levels of a **linear polyatomic** molecule are the same as for a diatomic molecule and characterized by one rotational constant and one rotational quantum number.

Non-linear molecules in which the moments of inertia about two axes are the same but different from the third are called **symmetric top rotors** since they have one axis of rotational symmetry. Example molecules are $NH_3$, $CH_3Cl$ and

$PCl_5$. The energy levels of a symmetric rotor are characterized by two rotational quantum numbers $J$ and $K$:

$$E_{J,K} = BJ(J + 1) + (A - B)K^2 \qquad J = 0,1,2,... \qquad K = J, J - 1,...,-J$$

The rotational constants $A$ and $B$ correspond to the moments of inertia parallel and perpendicular to the rotational symmetry axis of the molecule, respectively, and are given (in wavenumber units) by:

$$A = \frac{\hbar}{4\pi I_\parallel c} \qquad B = \frac{\hbar}{4\pi I_\perp c}$$

The quantum number $K$ indicates the extent of rotation about the symmetry axis. When $K = 0$ the molecule is rotating end-over-end only. Energy levels with $K > 0$ are doubly **degenerate** ($K$ appears as a squared term in the energy level expression) and correspond to states of clockwise or anticlockwise rotation about the symmetry axis.

In the most general case, a polyatomic molecule has three different moments of inertia (an **asymmetric rotor**) and the expressions for the rotational energy levels of these molecules are complex.

**Microwave rotational spectroscopy: selection rules and transitions**

For a molecule to interact with an electromagnetic field, and undergo a transition between two energy levels, requires an **electric dipole** to be associated with the motion giving rise to the energy levels. Since rotational motion does not change the electric dipole of a molecule, the **physical selection rule** for rotational energy transitions is:

*the molecule must possess a permanent electric dipole (i.e. the molecule must be polar)*

Symmetric molecules with a center of inversion, for example **homonuclear diatomics** ($H_2$, $O_2$) and **symmetric polyatomic molecules** ($CO_2$, $CH_4$ and $SF_6$) do not have permanent electric dipoles and therefore do not give rise to rotational spectra. **Heteronuclear diatomics** (e.g. HCl) and polyatomic molecules with no center of inversion (e.g. $NH_3$) do have rotational spectra.

The **specific selection rules** summarizing allowed transitions between rotational energy levels are:

$$\Delta J = \pm 1 \qquad \Delta K = 0$$

For transitions between energy levels with quantum numbers $J$ and $J+1$ the energy change is:

$$\Delta E = E_{J+1} - E_J = B(J + 1)(J + 2) - BJ(J + 1) = 2B(J + 1)$$

i.e. the energies of allowed rotational transitions are $2B$, $4B$, $6B$,.... Therefore the rotational spectra of a polar linear molecule and a polar symmetric top molecule consist of a series of lines at frequencies separated by energy $2B$ (*Fig. 1*). Rotational spectroscopy is often called **microwave spectroscopy** because values of $B$ are such that the energies of rotational transitions correspond to the microwave region of the spectrum. For example, $B = 1.921$ cm$^{-1}$ for carbon monoxide, so the $2B$ transition in the CO rotational spectrum occurs at a wavelength of 2.6 mm.

Microwave spectroscopy is useful for determining bond lengths from the moments of inertia derived from the separation of the lines in a rotational spectrum.

**Rotational intensities**

The **intensity** of a rotational transition is proportional to the population of the initial rotational energy level of the transition (see Topic I2) which depends on the **Boltzmann partition law** for that energy, $e^{-\frac{E_J}{k_BT}}$, and, the **degeneracy** of the level. The angular momentum of rotation is quantized, and each rotational level has a degeneracy of $2J + 1$ corresponding to the allowed orientations of the rotational angular momentum vector with respect to an external axis. Therefore the total relative population of a rotational level of a linear rigid rotor is

$$\propto (2J+1)e^{-\frac{BJ(J+1)}{k_BT}}.$$

The degeneracy factor causes the intensity of rotational transitions to increase linearly with $J$, whereas the Boltzmann factor causes the intensity to decrease exponentially with $J$. The net effect is a rotational spectrum with an intensity distribution that passes through a maximum (*Fig. 2*). The value of $J_{max}$ with the maximum population (and therefore giving rise to the transition of maximum intensity) is obtained by differentiating the above expression and setting this equal to zero:

$$J_{max} = \sqrt{\frac{k_BT}{2B}} - \frac{1}{2}$$

and is dependent on the temperature and the magnitude of $B$. (Note that the units of $B$ must be the same as the units of $k_BT$ when substituting numerical values.)

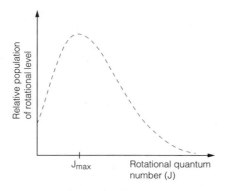

*Fig. 2.* Relative population of the rotational energy levels of a linear molecule as function of rotational quantum number J.

**Rotational Raman spectroscopy: selection rules and transitions**

In **Raman spectroscopy** (see Topic I1) an intense beam of monochromatic light is directed at the sample (typically **laser** light in the visible or ultraviolet) and the frequency of the scattered light is analyzed. Rotational Raman spectroscopy occurs when the incident photon interacts with the rotational energy of the molecule. Excitation of rotations in the molecule reduces the energy of the scattered photon and leads to scattered light of lower frequency than the incident radiation (**Stokes lines**). Conversely, transfer of rotational energy from the molecule to the incoming photon leads to scattered light of higher frequency (**anti-Stokes lines**).

The overall **physical selection rule** for rotational Raman spectra is that

*the **polarizability** of the molecule must be anisotropic.*

The polarizability of a molecule is a measure of the extent to which an applied electric field induces an **electric dipole** (Topic H6). The anisotropy is the variation of the polarizability with the orientation of the molecule relative to the electromagnetic field of the radiation. Entirely spherically symmetric molecules such as tetrahedral $CH_4$ or octahedral $SF_6$ have the same polarizability regardless of orientation so these molecules are rotationally **Raman inactive**. All non-spherically symmetric molecules are rotationally **Raman active**.

The **specific selection rules** for allowed rotational Raman transitions of linear molecules are:

$$\Delta J = +2 \text{ (Stokes lines)} \qquad \Delta J = -2 \text{ (anti-Stokes lines)}$$

so the energies corresponding to allowed transitions are given by:

$$\Delta E = E_{J+2} - E_J = B(J + 2)(J + 3) - BJ(J + 1) = 2B(2J + 3)$$

Therefore the rotational Raman spectrum consists of Stokes lines at energies $6B$, $10B$, $14B$,... less than the energy of the excitation line and anti-Stokes lines at energies $6B$, $10B$, $14B$,... greater than the energy of the excitation line (*Fig. 3*).

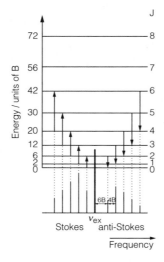

*Fig. 3.    The rotational Raman spectrum of a linear molecule.*

Rotational Raman spectroscopy permits determination of the rotational constant $B$ for non-polar **homonuclear diatomics** such as $H_2$ which do not give rise to microwave rotational spectra.

# I4 VIBRATIONAL SPECTROSCOPY

---

## Key Notes

**The harmonic oscillator**

In a harmonic oscillator the magnitude of the restoring force is directly proportional to the displacement from equilibrium and the resulting potential energy is proportional to the square of the displacement. At low vibrational energies, the vibration of a molecular bond can be approximated to that of a harmonic oscillator.

**Vibrational energy levels**

The energy of molecular vibration for the potential energy of a harmonic oscillator is quantized according to $E_v = (v + \frac{1}{2})hv$, where $v$ is the vibrational quantum number and $v = \frac{1}{2\pi}\sqrt{\frac{k}{\mu}}$ is the frequency of the oscillator. The parameters $k$ and $\mu$ are the force constant and the reduced mass of the vibration.

**Vibrational selection rules and transitions**

A molecular vibration is active in absorption or emission spectroscopy only if the electric dipole of the molecule changes during the vibration. The specific selection rule requires $\Delta v = \pm 1$. All allowed transitions of a harmonic oscillator of frequency, $v$, have energy $hv$.

**Related topics**

The wave nature of matter (G4)
General features of spectroscopy (I1)

Practical aspects of spectroscopy (I2)
Applied vibrational spectroscopy (I5)

---

**The harmonic oscillator**

The general shape of the potential energy curve for the stretching or bending of a molecular bond is shown in *Fig. 1*. The bottom of the potential energy well occurs at the equilibrium bond length $R_e$. The potential energy rises steeply for bond lengths $R < R_e$ because of strong repulsion between the positively-charged nuclei of the atoms at each end of the bond (Topic H6.). Potential energy rises as the atoms move further apart because of molecular bond distortion. Complete dissociation of the bond occurs as $R \rightarrow \infty$.

Near the bottom of the well, for small perturbations from $R_e$, the restoring force experienced by the atoms can be assumed to be directly proportional to the bond length displacement:

$$\text{restoring force} = -k(R - R_e)$$

where $k$ is called the **force constant** and is a measure of the strength of the bond. This is the **harmonic oscillator** approximation and gives a parabola for the potential energy curve, $V = \frac{1}{2}k(R - R_e)^2$. If displaced from equilibrium, the bond length undergoes the familiar sinusoidal harmonic oscillations, like two masses connected by a spring, or a clock pendulum.

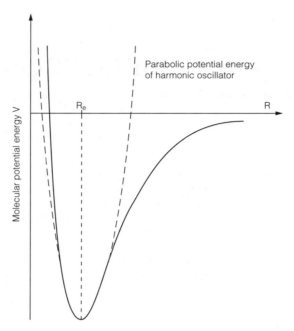

Fig. 1.   The approximation of a parabolic harmonic oscillator potential energy curve (– – –) to the molecular potential energy curve (———).

At high molecular vibrational energies the parabolic harmonic oscillator is a poor approximation of bond vibration and is replaced by a more sophisticated potential energy curve that includes the bond dissociation energy.

**Vibrational energy levels**

Vibrational energy, like all other molecular energy, is **quantized** (see Topic G4). The permitted values of vibrational energy are obtained by solving the **Schrödinger equation** for the motion of two atoms possessing the **harmonic oscillator** potential energy $V = \frac{1}{2}k(R - R_e)^2$. The allowed energy levels are:

$$E_v = \left(v + \frac{1}{2}\right)\hbar\omega \qquad v = 0,1,2,\ldots$$

where $\omega = \sqrt{k/\mu}$ is the circular frequency of the oscillator (in units of radians per second), and $\mu = m_1m_2/(m_1 + m_2)$ is the **reduced mass** of the two atoms. The circular frequency is related to the frequency $v$ (in units of $s^{-1}$, or hertz) by $\omega = 2\pi v$, so the permitted vibrational energy levels can also be written as:

$$E_v = \left(v + \frac{1}{2}\right)hv \qquad v = 0,1,2,\ldots$$

The integer $v$ is called the **vibrational quantum number**. The energy levels of a harmonic oscillator are evenly spaced with separation $hv$ (or $\hbar\omega$), as shown in Fig. 2.

Note that the magnitude of the vibrational energy levels depend on the reduced mass of the molecule, not the total mass. If one mass greatly exceeds the other (e.g. $m_2 \gg m_1$) the reduced mass is approximately equal to the lighter mass, $\mu \approx m_1$. In effect, the center of gravity is so close to the heavy mass that the light mass vibrates relative to a stationary anchor, e.g. the H atom in HI.

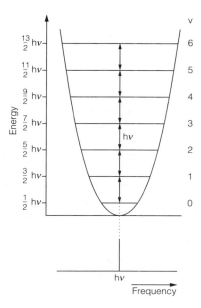

*Fig. 2.   The energy levels and allowed transitions of a harmonic oscillator.*

Bonds with greater frequency of oscillation (i.e. a strong bond and/or one with low reduced mass) have more widely separated vibrational energy levels. The lowest vibrational energy the molecule can possess corresponds to the lowest allowed vibrational quantum number ($v = 0$) and is $E_v = \frac{1}{2}hv$. This intrinsic **zero point energy** means that the atoms of a molecular vibration can never completely come to rest at their equilibrium bond separation.

**Vibrational selection rules and transitions**

Vibrational transitions only give rise to an observable spectrum in absorption or emission if the vibration interacts with electromagnetic radiation. Therefore the **physical selection rule** for vibrational spectra requires that

*the electric dipole of the molecule changes during the vibration.*

The molecule does not need to have a permanent electric dipole since the selection rule only requires a change in electric dipole during the vibration. Molecular vibrations which give rise to observable vibrational spectra are called **infrared active** modes. The vibrations of all **heteronuclear diatomics** are infrared active because the magnitude of the permanent electric dipole changes as the bond length changes during vibration. Conversely, all **homonuclear diatomics** (e.g. $O_2$, $N_2$) are **infrared inactive** because they have no permanent dipole and none arises during vibration. The polyatomic molecule $CO_2$ has no permanent dipole but its asymmetric stretching vibration and two bending vibrations give rise to oscillating electric dipoles and are therefore infrared active (see Topic I5). The symmetric stretching vibration of $CO_2$ is infrared inactive.

The **specific selection rule** for the allowed transitions between vibrational energy levels of a harmonic oscillator is:

$$\Delta v = \pm 1 \text{ only}$$

The positive value corresponds to absorption of energy from a lower to higher energy level, the negative value to emission.

Applying the selection rule, the energy of the transition between vibrational states with quantum numbers $v + 1$ and $v$ is:

$$\Delta E = \left(v + \frac{3}{2}\right)hv - \left(v + \frac{1}{2}\right)hv = hv$$

Since the transition energy is independent of quantum number, all transitions associated with a particular harmonic molecular vibration occur at a single frequency $v = 1/(2\pi)\ \sqrt{k/\mu}$, as shown in *Fig. 2*. Molecules with strong bonds (large $k$) between atoms of low masses (small $\mu$) have high vibrational frequencies. **Bending vibrations** are generally less stiff than **stretching vibrations**, so tend to occur at lower frequencies in the spectrum. At room temperature, application of the **Boltzmann distribution law** shows that almost all molecules are in their vibrational ground states and since all vibrational energy levels are singly **degenerate** (in contrast to rotational energy levels, Topic I3) the dominant spectral transition in absorption arises from $v = 0$ to $v = 1$.

A molecule can undergo a change in rotational energy level at the same time as it undergoes the change in vibrational energy level, subject to the appropriate rotational selection rules. It is often possible to resolve the rotational fine structure on either side of the position of the vibrational transition (similar in appearance to a pure rotational spectrum) in high resolution infrared spectroscopy of small molecules with large rotational constants.

# I5 APPLIED VIBRATIONAL SPECTROSCOPY

## Key Notes

**Infrared spectroscopy**

Transitions between molecular vibrational energy levels are associated with absorption or emission of radiation in the infrared portion of the electromagnetic spectrum, 100–10 000 cm⁻¹. Infrared spectroscopy is widely used to identify molecular vibrations characteristic of specific types of bonds.

**Polyatomic normal modes**

A normal mode is a collective vibrational displacement in which all atoms move in phase. For small displacements, normal modes are independent of each other, and can often be identified with particular types of bond vibration, for example C–H or C=O stretches. A non-linear molecule of $N$ atoms has $3N - 6$ normal modes; a linear molecule has $3N - 5$ modes.

**Vibrational Raman spectroscopy: selection rules and transitions**

Vibrational Raman spectroscopy requires a change in the polarizability of the molecule during the vibration. The specific selection rule allows only transitions of $\Delta v = \pm 1$. The Stokes and anti-Stokes vibrational transitions occur at frequencies $v_{ex} - v$ and $v_{ex} + v$, respectively, where $v$ is the frequency of the vibrational oscillator and $v_{ex}$ is the frequency of the incident excitation radiation.

**Rule of mutual exclusion**

The normal mode of a molecule with a center of symmetry cannot be both infrared and Raman active (and may be neither).

**Related topics**

General features of spectroscopy (I1)

Practical aspects of spectroscopy (I2)
Vibrational spectroscopy (I4)

---

**Infrared spectroscopy**

**Allowed transitions** between two **harmonic oscillator** vibrational energy levels requires **electromagnetic radiation** of the same frequency as the bond vibration, $v = 1/(2\pi) \sqrt{k/\mu}$, where $k$ is the force constant of the bond and $\mu$ is the reduced mass (see Topic I5). Example vibrational data for a number of diatomic molecules containing atoms of specific isotopes are given in *Table 1*.

Molecular vibrational frequencies generally lie in the **infrared** region of the electromagnetic spectrum ($3 \times 10^{12} - 3 \times 10^{14}$ Hz, or 100–10 000 cm⁻¹). A typical infrared absorption spectrometer records a spectrum in the frequency range 400–4000 cm⁻¹. Absorption at **characteristic frequencies** can often be ascribed to **polyatomic normal modes** of vibration of individual groups of atoms in the molecule. The characteristic frequencies of some of these vibrational modes are listed in *Table 2* and are useful for the identification of particular types of bond in an unknown compound. Absorptions from the remaining normal modes usually occur in the lower frequency **fingerprint region** ($< \sim 1200$ cm⁻¹).

Table 1.   Bond vibrational data for specific diatomic molecules

| Molecule | Force constant / N m$^{-1}$ | Reduced mass / amu | Vibration frequency / Hz | Vibration frequency / cm$^{-1}$ |
|---|---|---|---|---|
| $^1H^{19}F$ | 966 | 0.9570 | $1.241 \times 10^{14}$ | 4139 |
| $^1H^{35}Cl$ | 516 | 0.9796 | $8.964 \times 10^{13}$ | 2990 |
| $^1H^{81}Br$ | 412 | 0.9954 | $7.946 \times 10^{13}$ | 2649 |
| $^1H^{127}I$ | 314 | 0.9999 | $6.921 \times 10^{13}$ | 2308 |
| $^{12}C^{16}O$ | 1902 | 6.856 | $6.505 \times 10^{13}$ | 2170 |
| $^{14}N^{16}O$ | 1595 | 7.466 | $5.798 \times 10^{13}$ | 1904 |

Table 2.   Characteristic vibrational frequencies

| Vibration | Frequency / cm$^{-1}$ |
|---|---|
| Hydrogen bonds | 3200 – 3570 |
| O–H stretch | 3600 – 3650 |
| N–H stretch | 3200 – 3500 |
| =C–H stretch | 3000 – 3100 |
| C–H stretch | 2850 – 2970 |
| C≡N stretch | 2200 – 2270 |
| C≡C stretch | 2150 – 2260 |
| C=O stretch | 1650 – 1780 |
| C=C stretch | 1620 – 1680 |
| C–H bend | 1360 – 1470 |

## Polyatomic normal modes

A diatomic molecule possesses only one **mode of vibration**, the stretching and compression of the bond between the two atoms. The number of distinct modes of vibration in a non-linear polyatomic molecule containing $N > 2$ atoms is $3N - 6$. (To specify the displacement of each of $N$ atoms in three dimensions requires a total of $3N$ coordinates. Three of these coordinates specify the position of the center of mass of the molecule, and therefore correspond to the translational modes of the molecule, and three coordinates specify the orientation in space of the molecule, and therefore correspond to the rotational modes of the molecule.) A linear molecule of $N$ atoms possesses $3N - 5$ vibrational modes since only two angles are required to specify the orientation in space of a linear molecule.

The number of vibrational modes increases rapidly with the size of the molecule. For example, $H_2O$ is a non-linear triatomic molecule and has three modes of vibration, $CO_2$ is a linear triatomic molecule and has four modes of vibration, whereas benzene, $C_6H_6$, has 30 modes of vibration.

It is easier to visualize the vibrational modes of a polyatomic molecule when particular combinations of bond stretches or bends are considered together. These collective vibrational displacements, in which the atoms all move in phase and with the same frequency, are called **normal modes**. Exactly $3N - 6$, or $3N - 5$, independent normal modes of molecular vibration can be derived for non-linear, or linear, polyatomic molecules, respectively. Each normal mode behaves like a harmonic oscillator with a reduced mass, $\mu$, and force constant, $k$, that depend on which atoms and bonds contribute to the vibration (*Table 2*).

*Figure 1* illustrates the normal modes of vibration of $H_2O$ and $CO_2$. The bending vibration of $CO_2$ is doubly **degenerate** since the bending motion can also be drawn perpendicular to the plane of the paper in *Fig. 1b*. The degeneracy accounts for the required additional vibrational mode of linear

**(a)**

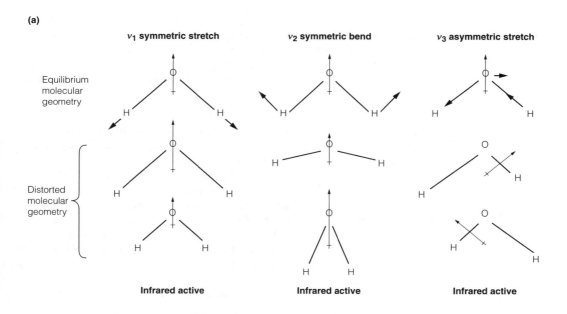

**(b)**

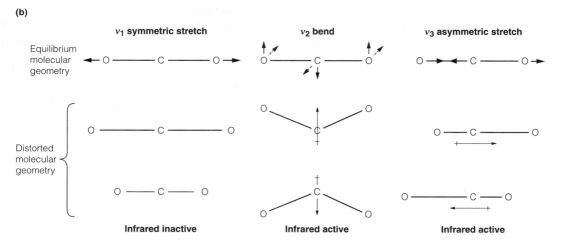

*Fig. 1.    Vibrational normal modes of (a) H₂O and (b) CO₂. The variation of electric dipole with vibrational distortion is also shown.*

molecules. Also shown in *Fig. 1* are the changes in **electric dipole** associated with the vibrations. Only vibrations which cause the electric dipole to oscillate are **infrared active** (see Topic I4). All vibrational modes of H₂O cause the electric dipole to vary (*Fig. 1a*) and so all modes are infrared active. The asymmetric stretching and bending modes of CO₂ are also infrared active, but the symmetric stretch leaves the dipole moment unchanged (at zero) so this mode is infrared inactive (*Fig. 1b*).

**Vibrational Raman spectroscopy: selection rules and transitions**

Raman spectroscopy occurs when a photon of incident radiation loses or gains energy in an interaction with a molecule (see Topic I1). The incident radiation must be monochromatic and intense and is typically provided by a **laser**.

The magnitude of the allowed energy exchange for **vibrational Raman spectroscopy** is determined by the same **specific selection rule** as for **infrared spectroscopy**:

$$\Delta v = \pm 1$$

Therefore vibrational **Stokes** and **anti-Stokes** radiation occur at frequencies $v_{ex} - v$ and $v_{ex} + v$, respectively, where $v_{ex}$ is the frequency of the incident excitation radiation. Anti-Stokes scattering can only arise if the molecule is already in an excited state. Since the proportion of molecules in vibrationally excited states is considerably smaller than in the ground state (determined by the **Boltzmann distribution** law), anti-Stokes transitions are much less intense than Stokes transitions. For small molecules, which have widely spaced vibrational and rotational energy levels, it may be possible to resolve rotational fine structure around the Stokes and anti-Stokes vibrational lines arising from the simultaneous loss and gain of rotational as well as vibrational energy in the scattering interaction.

In addition, a vibrational Raman line only occurs if

*the **polarizability** of the molecule changes during the vibration.*

The polarizability of a molecule is a measure of the extent to which an applied electric field, such as a photon of electromagnetic radiation, can induce an **electric dipole** (Topic H6). It is determined, in part, by the distribution of electron density in the molecular orbitals. A mode is **Raman active** if the vibration causes a change in either the magnitude or the three-dimensional shape of the polarizability. Both **homonuclear** and **heteronuclear diatomic** molecules swell and contract during vibration and the electron density changes non-symmetrically between the two extremes of displacement. Therefore all homonuclear and heteronuclear diatomics have Raman active vibrations, in contrast to infrared vibrational spectroscopy, in which homonuclear diatomic vibrations are inactive. The symmetric stretch of $CO_2$ is likewise Raman active. The asymmetric stretching and bending modes of $CO_2$ are not Raman active because the electron density changes symmetrically between the extremes of vibrational displacement in each mode and hence polarizability does not vary with small displacements from equilibrium.

**Rule of mutual exclusion**

The infrared and Raman activities of $CO_2$ vibrations are summarized in *Table 3*. The data demonstrate the **rule of mutual exclusion** for vibrational spectroscopy:

*If a molecule has a center of symmetry then no vibration can be both Raman and infrared active. (A mode may be inactive in both.) If there is no center of symmetry then some, but not necessarily all, vibrations may be both Raman and infrared active.*

The three vibrational modes of $H_2O$ (no center of symmetry) are all both Raman and infrared active.

*Table 3. Raman and infrared vibrational mode activities of $CO_2$*

| Mode of vibration | Raman | Infrared |
|---|---|---|
| $v_1$ (symmetric stretch) | Active | Inactive |
| $v_2$ (bend) | Inactive | Active |
| $v_1$ (asymmetric stretch) | Inactive | Active |

# I6 ELECTRONIC SPECTROSCOPY

## Key Notes

**UV/visible spectroscopy**

The separation of energy levels arising from different configurations of electrons in atomic and molecular orbitals usually corresponds to radiation in the visible and ultraviolet regions of the electromagnetic spectrum (wavelengths between 700 and ~100 nm).

**Franck–Condon principle**

Nuclei are sufficiently more massive than electrons that an electronic transition (electron rearrangement) occurs so fast that the nuclei do not alter their relative positions during the transition.

**Types of electronic transition**

The excitation of an electron from the $\pi$ bonding orbital of a C=C bond to the $\pi^*$ antibonding orbital is called a $\pi$–$\pi^*$ transition. The excitation of one of the lone pair electrons of the O atom in a C=O bond to the $\pi^*$ antibonding orbital is called an n–$\pi^*$ transition. The energy of $\pi$–$\pi^*$ and n–$\pi^*$ transitions shifts to longer wavelength radiation as conjugation of the C=C and C=O bonds increases. A charge transfer transition involves electron movement between the $d$ orbital of a metal atom and a ligand.

**Fluorescence and phosphorescence**

Fluorescence is the emission of radiation directly following absorption of excitation radiation. It is usually shifted to frequencies lower than the absorption because some vibrational excitation is lost in molecular collisions. Phosphorescence is the slow emission of radiation after absorption ceases and usually emanates from a triplet state accessed by spin-forbidden intersystem crossing from the initial excited singlet state.

**Photoelectron spectroscopy**

A photoelectron spectrum is obtained by measuring the kinetic energies of electrons emitted from a molecule following absorption of high energy (ultraviolet or X-ray) monochromatic radiation. The difference between the energy of the incident photon and the kinetic energy yields the energy of the orbital from which the electron was ejected.

**Related topics**

Valence bond theory (H2)
Molecular orbital theory of
   diatomic molecules I (H3)
Molecular orbital theory of
   diatomic molecules II (H4)

General features of spectroscopy
   (I1)
Practical aspects of spectroscopy
   (I2)
Photochemistry in the real world
   (I7)

**UV/visible spectroscopy**

Electronic energy levels are more widely separated than vibrational or rotational energy levels because considerably more energy is needed to change the distribution (i.e. **configuration**) of electrons in atomic or molecular orbitals than to change the energy of vibration or rotation. Consequently, **electronic spectroscopy** is usually associated with the visible or ultraviolet regions of the spectrum. The colors of many objects, for example vegetation, flowers, minerals, paints and dyes, are all due to **transitions** of electrons from one molecular orbital to another (see Topic I7).

The energy of ultraviolet photons is comparable with the strengths of many chemical bonds, so in some instances the absorption of light may lead to complete bond dissociation. The breakage of bonds in DNA by absorption of solar ultraviolet radiation is one factor in the formation of skin cancer from exposure to sunlight.

Atomic electronic spectra are described in Topic G7.

**Franck-Condon principle**

The different electronic states of a molecule are often associated with different shapes of the molecule because the different electron distribution around the molecule changes the electrostatic Coulombic forces that maintain the nuclei in specific relative positions. Since nuclei are considerably more massive than electrons, the **Franck-Condon principle** states that:

> *an electronic transition takes place sufficiently rapidly that the nuclei do not change their internuclear positions during the transition.*

Consequently, when energy is absorbed in an **electronic transition**, the nuclei suddenly find themselves in a new force field and at positions which are not in equilibrium for the new electronic state. This is shown schematically in *Fig. 1*, in which an electronic absorption from the ground state appears as a vertical line because of the Franck-Condon principle. The internuclear separation of the ground state becomes a **turning point**, the extent of maximum displacement, in a vibration of the excited state.

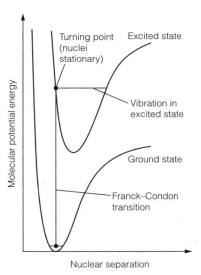

*Fig. 1.   Illustration of the Franck-Condon principle for vertical electronic transitions.*

The vertical transition has the greatest **transition probability** but transitions to nearby **vibrational levels** also occur with lower intensity. Therefore, instead of an electronic absorption occurring at a single, sharp line, electronic absorption consists of many lines each corresponding to the stimulation of different vibrations in the upper state. This **vibrational structure** (or **progression**) of an electronic transition can be resolved for small molecules in the gas-phase, but in a liquid or

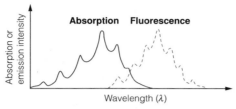

*Fig. 2. The relationship between the broad electronic absorption and fluorescence bands of liquids and solids.*

solid **collisional broadening** of the transitions cause the lines to merge together and the electronic absorption spectrum is often a broad band with limited structure (*Fig. 2*). The Franck-Condon principle also applies to downward transitions and accounts for the vibrational structure of a **fluorescence** spectrum.

**Types of electronic transition**

**Electronic transitions** arise from different types of electron rearrangement in the molecule, or groups of atoms in a molecule. A group of atoms that gives rise to a characteristic optical absorption is called a **chromophore**.

The C=O and C=C double bonds of organic molecules are common chromophores. The excitation of a bonding π electron in a C=C bond into an antibonding π* orbital by absorption of a photon is called a **π–π\* transition**. For an **unconjugated** double bond the energy of this transition corresponds to absorption of ultraviolet light at about 180 nm. When several double bonds form a **conjugated** chain, the energies of the extended π and π* orbitals lie closer together and the absorption transition shifts into the visible.

A similar, but weaker, electronic transition occurs in a carbonyl (C=O) group where one of the lone pair electrons on the O atom is excited into the antibonding π* molecular orbital of the carbonyl group. The absorption that promotes this **n–π\* transition** occurs at about 300 nm in the near ultraviolet. Carbonyl groups can also conjugate with C=C bonds and this again shifts the absorption towards the visible. The colors of many natural objects and synthetic dyes are due to π–π* and n–π* absorptions in conjugated systems, e.g. the carotene compounds responsible for the reds and yellows in vegetation.

Another common type of electronic transition, responsible for the intense color of many transition metal complexes and inorganic pigments, is a **charge-transfer** transition. In these transitions an electron transfers from the *d* orbitals of the metal to one of the ligands, or *vice versa*. An example of charge-transfer occurs in the permanganate ion, $MnO_4^-$. Absorption in the 420–700 nm range (responsible for the intense violet color) is associated with the redistribution of charge accompanying an electron transfer from an O atom to the Mn atom.

**Fluorescence and phosphorescence**

All electronically excited states have a finite lifetime. In most cases, particularly for large molecules in solids and liquids, the energy of excitation is dissipated into the disordered thermal motion of its surroundings. However, a molecule may also lose energy by **radiative decay,** with the emission of a photon as the electron transfers back into its lower energy orbital. There are two modes of radiative decay:

(i) **fluorescence**: the rapid spontaneous emission of radiation immediately following absorption of the excitation radiation;

(ii) **phosphorescence**: the emission of radiation over much longer timescales (seconds or even hours) following absorption of the excitation radiation.

The delay in phosphorescence is a consequence of energy storage in an intermediate, temporary reservoir.

A **Jablonski diagram** (*Fig. 3*) illustrates the relationship between fluorescence and phosphorescence and a typical arrangement of molecular electronic and vibrational energy levels. Therefore, the absorption of radiation promotes the molecule from the ground electronic state, $S_0$, to vibrationally excited levels in an upper electronic state, $S_1$. Therefore, the absorption spectrum shows structure (if any) characteristic of the vibrations of the upper state (*Fig. 2*). The $S$ nomenclature stands for **singlet state** and refers to the fact that the ground states of most molecules contain **paired** electron spins ($\uparrow\downarrow$), which can adopt only one orientation with respect to an external magnetic field.

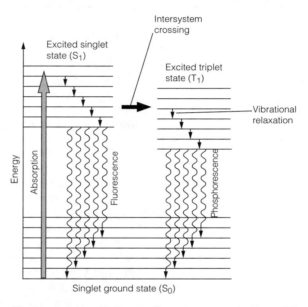

Fig. 3.   A Jablonski diagram illustrating energy levels participating in electronic absorption, fluorescence and phosphorescence.

Collisions of the excited molecule with surrounding molecules allow the excited state to lose its vibrational energy and sequentially step down the ladder of vibrational levels. The energy that the excited molecule needs to lose to return to the electronic ground state is usually too large for the surrounding molecules to accept, but if this energy is lost in a radiative transition, a fluorescence spectrum is produced. The observed fluorescence spectrum is shifted to lower frequency (longer wavelength) compared with the absorption spectrum (*Fig. 2*) because the fluorescence radiation is emitted after some of the molecules have already lost some vibrational energy (*Fig. 3*). The fluorescence spectrum therefore shows structure (if any) characteristic of the vibrations of the lower state.

A characteristic feature of a molecule that phosphoresces is that it possesses an excited **triplet electronic state**, $T_1$, of energy similar to the excited singlet state, $S_1$, and into which the excited singlet state can convert. In a triplet state two electrons in different orbitals have parallel spins ($\uparrow\uparrow$). Although normally forbidden, if a mechanism exists for converting paired ($\uparrow\downarrow$) electron spins into **unpaired** ($\uparrow\uparrow$) electron spins the excited $S_1$ state may undergo **intersystem** crossing into the $T_1$ state. (The usual mechanism for intersystem crossing is **spin-orbit coupling** in which the magnetic field from the nucleus of a heavy atom induces a neighboring electron to flip its spin orientation.) After intersystem crossing the molecule continues to step down the vibrational ladder of the $T_1$ state by loss of energy in collisions with surrounding molecules. The molecule cannot lose electronic energy by radiative transfer to the ground state because a triplet–singlet transition is forbidden. However, the transition is not entirely forbidden because the same mechanism that permits singlet to triplet intersystem crossing in the first place also breaks the selection rule so that the molecules are able to emit weak phosphorescence radiation on a longer timescale.

**Photoelectron spectroscopy**

The absorption of a photon of high enough energy may cause an electron to be ejected entirely from a molecule. In **photoelectron spectroscopy**, molecules are irradiated with high frequency, monochromatic light and the kinetic energy of the emitted photoelectrons is analyzed. The resulting photoelectron spectrum provides information on the energy levels of the orbitals from which the electrons were emitted. Conservation of energy dictates that if the incoming photon has frequency, $v$, and the ionization energy for the electron in an orbital is $I$, the kinetic energy of the emitted photoelectron is:

$$\frac{1}{2}mv^2 = hv - I$$

The kinetic energy of the electrons is determined from the strength of the electric or magnetic field required to bend their path into a detector. The slower the ejected electron, the lower in energy the molecular orbital from which it was ejected. Ultraviolet photoelectron spectroscopy provides information on the energy levels of the molecular orbitals of the **valence electrons** of molecules; X-ray photoelectron spectroscopy provides information on the energy levels of **core electrons**. If the apparatus has sufficient resolution of photoelectron kinetic energy, it may be possible to resolve fine structure in the photoelectron spectrum associated with the vibrational levels of the molecular cation formed by the ionization.

# 17 PHOTOCHEMISTRY IN THE REAL WORLD

## Key Notes

| | |
|---|---|
| **Atmospheric photochemistry** | The majority of chemical reaction sequences in the atmosphere are initiated by a photodissociation reaction in which absorption of a solar photon of visible or ultraviolet light promotes an electronic transition leading to bond dissociation and formation of reactive radicals. The photodissociation of $O_2$ leading to the formation of the ozone layer in the stratosphere is an example. |
| **Photosynthesis** | The energy to drive photosynthesis in plants is provided by the absorption of photons of visible light by molecules of chlorophyll. The excited electron produced in the electronic transition increases the redox potential of a neighboring electron acceptor molecule. Plants appear green because chlorophyll molecules strongly absorb the blue and red components of the visible spectrum whilst green is reflected. |
| **Vision** | The electronic transition induced in the retinal molecule by absorption of a photon of visible light causes the molecule to change configuration between *cis* and *trans* isomeric forms. The *trans* isomer of retinal is unable to retain its binding to its associated protein in the eye's retina and the dissociation triggers a nerve impulse to the brain. |
| **Chemiluminescence and bioluminescence** | Chemiluminescence is the emission of light from an electronically excited state produced by a chemical reaction (e.g. the blue emission from hydrocarbon radicals in a flame). Bioluminescence is a chemiluminescent reaction that occurs in a living organism. The light emitted by the luciferin reaction in a firefly is a familiar example. |
| **Related topic** | Electronic spectroscopy (I6) |

**Atmospheric photochemistry**

The chemistry of the atmosphere is initiated by reactions involving the absorption of light from the sun. The energy of visible and ultraviolet photons is sufficient to cause bond dissociation and the formation of radical photodissociation products. One of the most important photochemical processes in the atmosphere leads to the formation of the ozone ($O_3$) layer in the stratosphere, which is the region of the Earth's atmosphere extending between about 15 and 50 km above the surface.

Molecular oxygen, $O_2$, in the stratosphere absorbs solar ultraviolet photons of wavelength < 240 nm and dissociates into O atoms which rapidly combine with surrounding undissociated $O_2$ to form $O_3$:

$$O_2 + h\nu \rightarrow O + O$$

$$O + O_2 \rightarrow O_3$$

The ozone concentration is maintained at a **steady state** in the stratosphere by the photodissociation of $O_3$ itself back to an O atom and the reformation of diatomic $O_2$ by reaction of O with $O_3$:

$$O_3 + hv \rightarrow O_2 + O$$

$$O + O_3 \rightarrow O_2 + O_2$$

It is the absorption of solar near-ultraviolet radiation (280–380 nm) through the photodissociation of $O_3$ in the stratosphere that prevents these wavelengths from penetrating to the Earth's surface where they can cause biological damage.

**Photosynthesis**

The most important biological example of an electronic excitation induced by absorption of a photon is the transition that initiates the process of **photosynthesis** in plants. In broad terms, photosynthesis uses the energy of solar photons to drive a series of reactions which produce sugars and carbohydrates from carbon dioxide and water. The photons are absorbed by **chlorophyll** molecules (*Fig. 1*) located in the membranes of the chloroplast structures found in the cells of all green vegetation. A chlorophyll molecule consists of a metal atom surrounded by a **conjugated** ring system (a porphyrin) and has an absorption **transition** in the visible part of the electromagnetic spectrum.

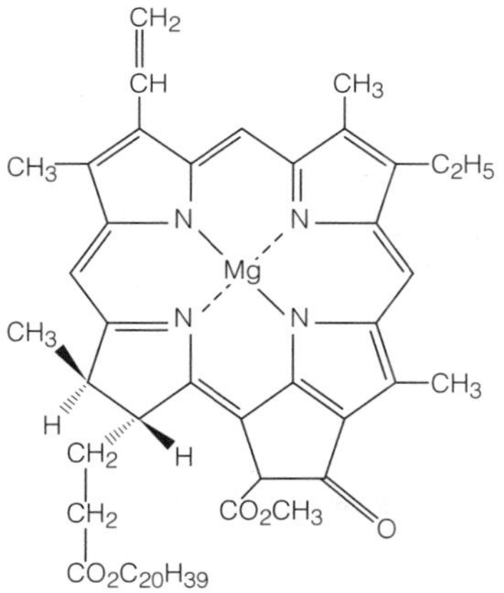

*Fig. 1.    The structure of the chlorophyll molecule.*

In an isolated chlorophyll molecule the energy of the excited electronic state dissipates as **fluorescence** or to thermal motion of surrounding molecules, but in the chloroplast the chlorophyll molecules are bound tightly by proteins to electron acceptors and electron donors. The collection of molecules is called a **photosystem**. The excited electron produced in the absorption process is passed rapidly to the electron acceptor before fluorescence can occur. The higher **redox potential** created in the electron acceptor supplies energy for the sugar formation reactions. The neighboring electron donor returns the chlorophyll molecule to its ground state, ready to absorb another photon.

The energy of a single photon of visible light is not sufficient to drive the whole sequence of sugar formation reactions and two slightly different photosystems operate in tandem in the chloroplasts. This is the origin of the green color of vegetation. One chlorophyll photosystem absorbs strongly at the red end of the visible spectrum and the other at the blue end, leaving green as the dominant color in the reflected light.

The theoretical maximum **quantum yield** of photosynthesis (the number of molecules of $CO_2$ converted to sugar as a ratio of the number of absorbed quanta) is $1/8$. In practice, fluorescence and non-electron transfer decay of excited chlorophyll molecules and loss of $CO_2$ by respiration reduces the quantum yield efficiency to around $1/15$, depending on local biochemical and environmental factors.

**Vision**

The mechanism of human vision is an example in which absorption of photons of visible light induces a $\pi$–$\pi^*$ **electronic transition** in a molecule containing conjugated C=C bonds (see Topic I6). The molecule, 11-*cis*-retinal (*Fig. 2*), combines with the protein opsin in the retina of the eye to form rhodopsin. The combination of the 11-*cis*-retinal chromophore with the protein shifts the absorption maximum of 11-*cis*-retinal into the visible.

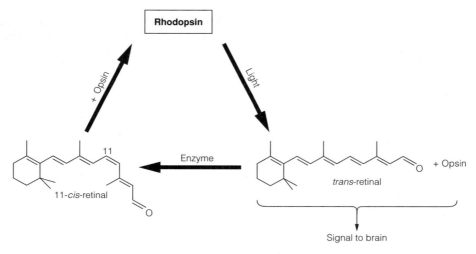

Fig. 2.   *The role of photo-excitation of retinal in the mechanism of vision.*

When a photon of visible light is absorbed by a molecule of rhodopsin, the double bond at position 11 in retinal is free to isomerize to the more stable *trans* configuration because the C=C bond in the excited electronic state is no longer torsionally rigid. However, the spatial interaction between *trans*-11-retinal and the opsin protein is unfavorable, so the rhodopsin molecule dissociates and this triggers a nerve impulse to the brain (*Fig. 2*). An enzyme promotes isomerization of *trans*-11-retinal back to *cis*-11-retinal, rhodopsin reforms and the visual cycle restarts.

**Chemilumine-
scence and
bioluminescence**

The emission of visible or ultraviolet light by a molecule promoted to an excited state by a chemical reaction, rather than by absorption of light, is called **chemiluminescence**. The species excited initially in the reaction may emit light itself on relaxation to the ground state or transfer its energy to another molecule which then emits.

One example of chemiluminescence is the blue light associated with combustion flames, e.g. a gas oven ring. The oxidation reactions of the hydrocarbon fuel produce transient CH and CHO radicals in electronically excited states which emit at discrete frequencies in the visible. An example of natural chemiluminescence is the light emission associated with the atmospheric aurora. Reactions between molecules in the upper atmosphere and high energy particles from the solar wind produce some atoms, molecules and ions in electronically excited states which emit visible fluorescence as they return to the ground state.

Chemiluminescence that takes place in living organisms is generally known as **bioluminescence**. A number of biological organisms (e.g. the firefly and many marine creatures inhabiting the darker depths of the ocean) emit light with a range of frequencies from ultraviolet to the red end of the visible spectrum. The light emitted by a firefly is the result of visible emission from an excited product derived from luciferin, and is one of the most efficient chemiluminescence systems known. The enzyme luciferase catalyzes oxidation of luciferin to an intermediate which loses $CO_2$ to form the excited product that is the source of the light emission (*Fig. 3*).

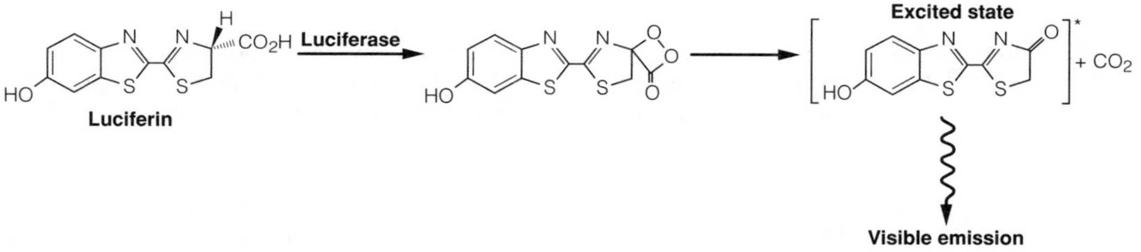

Fig. 3.    *The sequence of reactions leading to the bioluminescence produced by fireflies.*

# 18 MAGNETIC RESONANCE SPECTROSCOPY

---

## Key Notes

**Principles of nuclear magnetic resonance spectroscopy**
Nuclear magnetic resonance (NMR) spectroscopy is the resonant absorption of radiofrequency radiation when nuclei with non-zero spin angular momentum convert between spin states separated in energy by an applied magnetic field. NMR provides information on the different chemical environments of nuclei in a molecule. The magnitude of the absorption is proportional to the number of equivalent nuclei in the same environment.

**Chemical shift**
The frequency of nuclear magnetic resonance absorption depends on the local magnetic environment of the nuclei in the molecule and is reported as a chemical shift, $\delta$, relative to the resonance frequency of a reference standard.

**Fine structure**
The NMR resonance absorption due to a group of equivalent nuclei is split into fine structure if coupling occurs between their magnetic moment and those of neighboring nuclei in the molecule. The magnitude of the splitting is measured by a spin–spin coupling constant. A group of $N$ equivalent protons splits the absorption line of a nearby group into $N + 1$ lines with intensities given by the coefficients of the $(N + 1)$th binomial expansion.

**Electron spin resonance spectroscopy**
Electron spin resonance (ESR) is the resonance absorption of microwave radiation by unpaired electrons in a magnetic field. The technique provides information on the electronic structure of radicals, triplet states and $d$-metal complexes with unpaired electrons. An ESR absorption shows hyperfine structure if the electron couples with a neighboring magnetic nucleus.

**Related topic**
General features of spectroscopy (I1)

---

**Principles of nuclear magnetic resonance spectroscopy**

**Nuclear magnetic resonance (NMR)** spectroscopy provides information on the chemical environment of the nuclei in a molecule. Many atomic nuclei possess spin angular momentum which is quantified by a **nuclear spin quantum number**, $I$. The nuclear spin angular momentum may take $2I + 1$ different orientations relative to an arbitrary axis in space, each of which is distinguished by values of the quantum number, $m_I$,

$$m_I = I, (I - 1), \ldots, -I$$

The nuclear spin angular momentum quantum number can have a range of both integral and half-integral values, as well as zero. Values for some common nuclei are shown in *Table 1*.

*Table 1.    Nuclear spin quantum number and abundance for some common isotopes*

| Isotope | Natural abundance/% | Spin I |
|---------|---------------------|--------|
| $^{1}H$ | 99.98 | ½ |
| $^{2}H$ | 0.016 | 1 |
| $^{12}C$ | 98.99 | 0 |
| $^{13}C$ | 1.11 | ½ |
| $^{14}N$ | 99.64 | 1 |
| $^{16}O$ | 99.96 | 0 |
| $^{17}O$ | 0.037 | ⁵⁄₂ |
| $^{19}F$ | 100 | ½ |

A nucleus with non-zero spin behaves like a magnet. In the presence of a magnetic field, $B$ (units tesla, T), the **degeneracy** of nuclei with the $2I + 1$ possible orientations of nuclear spin angular momentum is removed. The states acquire different values of potential energy:

$$E = -Bm_I g_I \mu_N$$

where $g_I$ is a numerical **g-factor** characteristic of the nucleus (and determined experimentally), and $\mu_N = e\hbar/2m_p$ is the **nuclear magneton** with value $5.05 \times 10^{-27}$ J T$^{-1}$ ($m_p$ is the mass of the proton).

The two spin states ($m_I = ½$ and $m_I = -½$) of a hydrogen $^{1}H$ nucleus in a magnetic field (or any other nucleus with I = ½) are separated by an energy (*Fig. 1*).

$$\Delta E = \frac{1}{2}Bg_I\mu_N - \left(-\frac{1}{2}Bg_I\mu_N\right) = Bg_I\mu_N$$

The **Boltzmann distribution law** dictates that slightly more nuclei will be in the lower of the two energy states. Electromagnetic radiation of energy resonant with the energy separation induces transitions between the two spin states and is strongly absorbed. The resonance frequency, $\nu = Bg_I\mu_N/h$, is proportional to the strength of the magnetic field, and is in the radiofrequency region of the spectrum.

In the practical application of Fourier Transform NMR spectroscopy, the sample is subjected simultaneously to a magnetic field and pulses of radiofrequency

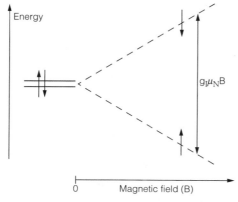

*Fig. 1.    The splitting of the energy levels of a spin ½ nucleus by an applied magnetic field.*

electromagnetic radiation, and the spectrum of absorbed frequencies at which the nuclei come into resonance are recorded. A useful feature of NMR spectroscopy is that the **intensity** of the NMR absorption is directly proportional to the number of nuclei giving rise to the transition. A larger magnetic field also increases the intensity of the absorption by increasing the energy separation and hence the population difference between the nuclear spin states.

Since possession of spin angular momentum is a necessary condition for an atomic nucleus to give an NMR signal, the fact that both $^{12}C$ and $^{16}O$ nuclei have zero spin (*Table 1*) considerably simplifies the interpretation of $^1H$-NMR spectra of organic compounds. NMR spectroscopy is now applied to many active nuclei, e.g. $^{13}C$.

**Chemical shift**

The exact radiofrequency of NMR absorption for a particular nucleus depends on both the strength of the applied external magnetic field and how this field is moderated by the local electronic structure in the molecule. The strength of the local perturbation, $\delta B$, is proportional to the strength of the applied field, $\delta B = \sigma B$, where $\sigma$ is called the **shielding constant**. The constant is positive or negative according to whether the perturbation is in the same or opposite direction to the applied field. The degree of perturbation depends on the particular electronic structure near the magnetic nucleus of interest, so different nuclei, even of the same element, undergo resonance absorption at different frequencies.

The frequencies of resonance absorption for particular nuclei are quantified in terms of their **chemical shift** from the frequency of a reference standard. The standard for proton ($^1H$) NMR spectroscopy is tetramethylsilane (TMS), $Si(CH_3)_4$, which has a single resonance absorption because the molecular symmetry ensures all protons are in equivalent chemical environments. Chemical shifts are reported as a $\delta$ **value** relative to the resonance frequency of the standard, $v^\ominus$:

$$\delta = \frac{v - v^\ominus}{v^\ominus}$$

in order that they are independent of the applied field. Values of $\delta$ are usually of the order $10^{-6}$ and are conventionally expressed as parts per million (ppm). Typical chemical shifts of protons in particular chemical environments are given in *Table 2*. Nuclei with values $\delta > 0$ are referred to as **deshielded**, i.e. the local magnetic field experienced by these nuclei is stronger than that experienced by the nuclei in the standard under the same conditions. Nearby electron-withdrawing substituents cause increased deshielding.

The NMR spectrum of ethanol ($CH_3CH_2OH$) is shown in *Fig. 2*. The three distinct chemical shifts indicate protons in three different types of environment. Since the intensity of an NMR signal is proportional to the number of equivalent nuclei giving the resonance, the integrated intensities of the three groups of lines are in the ratio 3:2:1 for the three $CH_3$ protons, two $CH_2$ protons and one OH proton, respectively. This quantitative property is a useful feature of NMR spectroscopy.

**Fine structure**

In addition to the effects of **deshielding**, the local magnetic field experienced by a particular nucleus (or by equivalent nuclei) is also influenced by the presence of other magnetic nuclei nearby and this creates **fine structure** in the corresponding resonance frequency in the NMR spectrum. *Fig. 2* shows the fine structure in the NMR spectrum of ethanol. Fine structure splitting (spin-splitting) is not observed between nuclei in equivalent chemical environments. The extent of

*Table 2. Typical chemical shifts for ¹H nuclear magnetic resonances. Uncertainties are in the range ± 0.5 ppm, except for values marked with ~ which vary more widely with sample conditions of concentration, pH and solvent*

| ¹H environment | $\delta$ (ppm) |
|---|---|
| Si(CH$_3$)$_4$ | 0 |
| R–CH$_3$ | 0.9 |
| –CH$_2$– | 1.4 |
| R–NH$_2$ | ~ 2 |
| R–CO$_2$–CH$_3$ | 2.1 |
| R–CO–CH$_3$ | 2.2 |
| –C≡CH | 2.2 |
| Ph–CH$_3$ | 2.3 |
| R–O–CH$_3$ | 3.3 |
| R–CO$_2$–CH$_3$ | 3.7 |
| R–OH | ~ 4 |
| –C=CH– | 5.1 |
| Ph–H | 7.3 |
| –CHO | ~ 10 |
| R–COOH | ~ 12 |

fine structure splitting is determined by the **spin–spin coupling constant**, $J$, between the two sets of non-equivalent magnetic nuclei. Spin–spin coupling constants are an intrinsic property of the molecule and independent of the strength of the applied magnetic field.

The local magnetic field experienced by a single proton A, coupled to a single proton B, is equal to the sum of the applied field, plus the perturbation due to

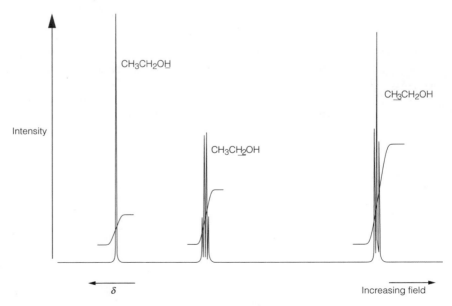

*Fig. 2. The NMR spectrum of ethanol. The protons giving rise to each group of lines are indicated by underlining. The cumulative integration of the intensity of each group of lines is also shown.*

deshielding, and plus or minus the perturbation due to nucleus B, depending on whether the magnetic moment of B is aligned with or against the applied field. These two latter possibilities are equally likely. Therefore, a doublet resonance of equal intensity is observed at frequencies $+\frac{1}{2}J_{A-B}$ and $-\frac{1}{2}J_{A-B}$ from the single resonance frequency that would be observed for A in the absence of coupling with B, where $J_{A-B}$ is the spin–spin coupling constant specific to A and B (*Fig. 3*). Since nucleus A has the same effect on nucleus B, the chemical shift for B is also split into a doublet with the same frequency separation $J_{A-B}$.

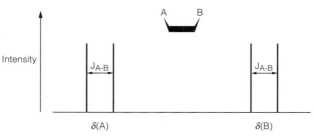

*Fig. 3.* The chemical shifts of two non-equivalent protons are split into doublets by spin–spin coupling.

If there are two equivalent protons B present, the chemical shift of proton A splits into a triplet of lines of separation $J_{A-B}$ (*Fig. 4*). The intensity ratio of the triplet is 1:2:1 because there is one combination in which both B magnetic moments are aligned with the applied field, two combinations in which one B magnetic moment is aligned with the field and one aligned against the field, and one combination in which both B magnetic moments are aligned against the field. The concept is readily extended; N equivalent protons B split the chemical shift of proton A into $N+1$ lines with intensity ratio given by the coefficients of the Binomial expansion to power $N+1$ or, equivalently, to the $(N+1)$th line of Pascal's triangle.

In general, spin–spin coupling in proton NMR spectra is only important between protons attached to adjacently bonded atoms. Couplings over larger numbers of bonds can be ignored.

In the ethanol NMR spectrum (*Fig. 2*) the three protons of the $CH_3$ group split the single resonance peak of the $CH_2$ protons into a 1:3:3:1 quartet. The two

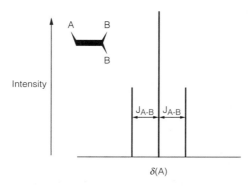

*Fig. 4.* The chemical shift of a proton splits into a 1:2:1 triplet when coupled with two equivalent protons.

protons of the $CH_2$ group split the resonance peak of the $CH_3$ protons into a 1:2:1 triplet. The coupling between the $CH_2$ protons and the OH proton is small and no additional splitting is observed.

**Electron spin resonance spectroscopy**

Electron spin resonance (ESR) spectroscopy is analogous to NMR spectroscopy. An electron possesses half-integral **spin angular momentum** which gives rise to two possible spin orientations distinguished by the quantum numbers $m_s = \frac{1}{2}$ and $m_s = -\frac{1}{2}$ (Topic G6). A strong magnetic field is used to remove the **degeneracy** of the two orientations of the electron magnetic moment and transitions between the two states are induced by the absorption of radiation resonant with the energy separation.

ESR differs from NMR in an important respect. The **Pauli exclusion principle** requires that whenever two electrons occupy one orbital their spins must be **paired**. It is therefore only possible to reorientate the spin of an electron if the electron is unpaired, and so ESR is restricted to species with an odd number of electrons (radicals, **triplet states**, and $d$-metal complexes). Closed shell species give no ESR absorption. The intensity of an ESR absorption is proportional to the concentration of the species with the unpaired electron present.

The frequency of the resonance absorption is obtained by analogy with NMR. The two electron spin states ($m_s = \pm\frac{1}{2}$) are separated by energy:

$$h\nu = \frac{1}{2}Bg\mu_B - \left(-\frac{1}{2}Bg\mu_B\right) = Bg\mu_B$$

where $B$ is the applied field strength, $g$ is the **$g$-factor** for a free electron and $\mu_B = e\hbar/2m_e$ is the **Bohr magneton** ($m_e$ is the mass of an electron). The value of the Bohr magneton ($9.3 \times 10^{-24}$ J T$^{-1}$) is 1837 times greater than the nuclear magneton because of the difference in mass between an electron and proton. Consequently, the energy separation of the two electron spin orientations is greater for a given applied field than for nuclear spin orientations and ESR resonance absorptions occur in the microwave rather than the radiofrequency region of the spectrum.

As with NMR spectroscopy, ESR signals can show **hyperfine structure** caused by the coupling of the electron with magnetic nuclei. If an electron interacts with just one nucleus of spin ½, then its ESR signal consists of two peaks of equal intensity (separated in frequency by the appropriate **coupling constant**), corresponding to the two possible orientations of the nucleus. The ESR signal for the $CH_3$ radical consists of four peaks in a 1:3:3:1 ratio because of the combinations in which an electron spin can couple to the spins of three equivalent protons.

ESR provides information on the electronic structure of radicals. The value of the coupling constant between the electron and the nucleus in a free hydrogen atom is 1420 MHz. If the value of the coupling constant of an electron to a hydrogen atom in a radical is $A$ MHz, then the spin population on that hydrogen atom is very approximately $A/1420$. Similarly, the coupling constant between a $2s$ electron and a $^{13}C$ nucleus is 3330 MHz, so the coupling constant of an electron in an $sp^3$ hybridized orbital is approximately one-quarter of this. Different values of coupling constants indicate different degrees of **hybridization**.

# APPENDIX – MATHEMATICAL RELATIONS

**Units**

Dimensional analysis allows the units of the quantities in an equation to predict the units of the answer which is obtained. The units in an equation are separated from the numerical part of the equation, and treated as variables which may be cancelled out or multiplied. If, for example, the kinetic energy of a ball is calculated from $E = \frac{1}{2} mv^2$, the units of mass, $m$, are kg, the units of speed, $v$, are m s$^{-1}$, hence the units of kinetic energy are kg (m s$^{-1}$)$^2$ =kg m$^2$ s$^{-2}$.

This also illustrates the fact that the units of energy (usually given as Joules, J) may also be expressed in terms of fundamental (cgs) units. The Joule is known as a **derived** unit. Common examples of derived units are given in *Table 1*.

Table 1.   Derived units and their equivalent fundamental units

| Quantity | Derived unit | Fundamental (cgs) units |
|---|---|---|
| Energy | J | m$^2$ kg s$^{-2}$ |
| Frequency | Hz | s$^{-1}$ |
| Force | N | m kg s$^{-2}$ |
| Pressure | Pa | m$^{-1}$ kg s$^{-2}$ |
| Charge | C | A s |
| Potential | V | m$^2$ kg s$^{-3}$ A$^{-1}$ |
| Resistance | Ω | m$^2$ kg s$^{-3}$ A$^{-2}$ |

**Approximations**

$A \pm x \approx A$        if $A \gg x$

$x/A \approx 0$        if $x \ll A$

$\ln(N!) = N\ln(N) - N$        for large values of N (Stirling's approximation)

$\ln(1+x) \approx x$        as $x \to 0$

$e^x \approx 1+x$        as $x \to 0$

**Sums, products and differences**

$\Delta x = $ (final value of $x$) $-$ (initial value of $x$)

$$\sum_{i=1}^{n} x_i = x_1 + x_2 + x_3 + \cdots + x_{n-1} + x_n$$

$$\prod_{i=1}^{n} x_i = x_1 x_2 x_3 (\ldots) x_{n-1} x_n$$

**Logarithms and exponentials**

If $y = a^x$, then $\log_a(y) = x$

a is the **base** of the log. Logs in base e are denoted as $\ln(y)$.

Mathematical definitions require that $x$ and $y$ must both be dimensionless.

In the absence of a subscript, $\log(x)$ usually implies $\log_{10}(x)$ and $\ln(x)$ denotes $\log_e(x)$

Basic relationships:

$\log (xy) = \log (x) + \log (y)$

$\log (x/y) = \log (x) - \log (y)$

$\log (x^y) = y \log (x)$

Converting from base $a$ to base $b$:

$$\log_a (x) = \log_a(b) \cdot \log_b (x)$$
$$\text{Hence, } \log_e (x) = 2.303 \log_{10} (x)$$

**Differentials**

In all cases, $a$ and $c$ are independent of $x$, i.e. $a \neq f(x)$ and $c \neq f(x)$.

Basic differential relationships:

$$\frac{d}{dx} x^n = nx^{n-1} \qquad \frac{d}{dx} \ln(ax) = \frac{1}{x} \qquad \frac{d}{dx} e^{ax} = ae^{ax}$$
$$\frac{d}{dx} \sin(ax) = a \cos(ax) \qquad \frac{d}{dx} \cos(ax) = -a \sin(ax)$$

Differentials of products, sums, and quotients ($f$, $g$, $u$, $v$ are all functions of $x$):

$$\frac{d}{dx}\big(f(x) + g(x)\big) = \frac{d}{dx} f(x) + \frac{d}{dx} g(x)$$
$$\frac{d}{dx}(uv) = u\frac{dv}{dx} + v\frac{du}{dx} \qquad \frac{d}{dx}\left(\frac{u}{v}\right) = \frac{v\dfrac{du}{dx} - u\dfrac{dv}{dx}}{v^2}$$

Fundamental relationship for $f(x, y)$:

$$\frac{df}{dx} = \frac{df}{dy} \cdot \frac{dy}{dx}$$

**Indefinite integrals**

$$\int \big(f(x) + g(x)\big)dx = \int f(x) + \int g(x)$$
$$\int ax^n dx = a \int x^n dx = \frac{ax^{n+1}}{n+1} + c \qquad \text{for } n \neq -1$$
$$\int \frac{dx}{x} = \ln(x) + c \qquad \text{for } x \neq 0$$

$c$ is not a function of $x$.

**Definite integrals**

$$\int_R^S ax^n dx = a \int_R^S x^n dx = \left[\frac{ax^{n+1}}{n+1} + c\right]_R^S = \left[\frac{ax^{n+1}}{n+1}\right]_R^S = \frac{aS^{n+1}}{n+1} - \frac{aR^{n+1}}{n+1} \qquad \text{for } n \neq -1$$
$$\int_a^b f(x)dx = -\int_b^a f(x)dx$$

**Factorials**

$N! = N.(N-1).(N-2).(N-3)\ldots3.2.1 \qquad \text{for } N \neq 0$
$0! = 1$

# FURTHER READING

## General reading

Atkins, P.W. (1998) *Physical Chemistry*, 6th edn. Oxford University Press, Oxford.
Atkins, P.W. (1996) *The Elements of Physical Chemistry*. 2nd edn, Oxford University Press, Oxford.
Mahan, B.M. and Myers, R.J. (1987) *University Chemistry*. Benjamin/Cummings Publishing Co., Menlo Park.
Laidler, K.J. (1978) *Physical Chemistry with Biological Applications*. Benjamin/Cummings, London.
Moore, W.J. (1983) *Basic Physical Chemistry*. Prentice-Hall.
Chang, R. (2000) *Physical Chemistry with Applications to Biological Systems*. Macmillan.

## More specific references

**Section A**    Tabor, D. (1979) *Gases, Liquids and Solids*. Cambridge University Press, Cambridge.
Hirschfelder, J.O. and Curtiss, C.F. (1992) *The Molecular Theory of Gases and Liquids*. Wiley, New York.
Barton, A.F.M. (1974) *The Dynamic Liquid State*. Longman, New York.
Murrell, J.N. and Boucher, E.A. (1982) *Properties of Liquids and Solutions*. John Wiley and Sons, New York.
Wells, A.F. (1984) *Structural Inorganic Chemistry*. Clarendon Press, Oxford.
West, A.R. (1990) *Solid State Chemistry and its Applications*. Wiley, London.

**Section B**    Smith, E.B. (1990) *Basic Chemical Thermodynamics*. Oxford University Press, Oxford.
McGlashan, M.L. (1979) *Chemical Thermodynamics*. Academic Press, London.
Caldin, E.F. (1961) *Chemical Thermodynamics*. Clarendon Press, Oxford.
Warn, J.R.W. (1971) *Concise Chemical Thermodynamics*. Van Nostrand Reinhold.
Fletcher, P. (1993) *Chemical Thermodynamics for Earth Scientists*. Longman, Harlow.

**Section C**    King, E.J. (1965) *The International Encylopædia of Physical Chemistry and Chemical Physics, vol 4. Acid-Base Equilibria*. Pergamon Press, Oxford.
Bell, R.P. (1969) *Acids and Bases*. Methuen, London.
Mattock, G. (1961) *pH Measurement and Titration*. Heywood and Co., London.
Blandamer, M.J. (1992) *Chemical Equilibria in Solution*. Ellis Horwood – Prentice Hall, Hemel Hempsted.
Denbigh, K. (1981) *The Principles of Chemical Equilibrium*. Cambridge University Press, Cambridge.
Bates, R.G. (1973) *Determination of pH*. Wiley, New York.

**Section D**    Fletcher, P. (1993) *Chemical Thermodynamics for Earth Scientists*. Longman, Harlow.
Murrell, J.N. and Boucher, E.A. (1982) *Properties of Liquids and Solutions*. John Wiley and Sons, New York.
Rowlinson, J.S. and Swinton, F.L. (1994) *Liquids and Liquid Mixtures*. Butterworth, London.
Alper, A. (1970) *Phase Diagrams* (3 vols). Academic Press, New York.

Stanley, H.E. (1971) *Introduction to Phase Transitions and Critical Phenomena.* Clarendon Press, Oxford.

Findlay, A., Cambell, A.N. and Smith, N.O. (1951) *Phase Rule and its Applications.* Cover Publications.

**Section E**

Hibbert, D.B. (1993) *Introduction to Electrochemistry.* Macmillan.

Robbins, J. (1972) *Ions in Solution (2) An Introduction to Electrochemistry.* Clarendon Press, Oxford.

Bard, A.J. and Faulkner, L.R. (1980) *Electrochemical Methods, Fundamentals and Applications.* John Wiley and Sons, Chichester.

Bazier, M.M. and Lund, H. (editors) (1983) *Organic Electrochemistry – an Introduction and a Guide.* Marcel Dekker, New York.

Compton, R.G. and Sanders, G.H.W. (1996) *Electrode Potentials.* Oxford University Press, Oxford.

Cantor, C.R. and Schimmel, P.R. (1980) *Biophysical Chemistry III. The Behaviour of Biological Macromolecules.* Freeman, San Francisco.

**Section F**

Pilling, M.J. and Seakins, P.W. (1995) *Reaction Kinetics.* Oxford University Press, Oxford.

Laidler, K.J. (1987) *Chemical Kinetics,* 3rd edn. HarperCollins, New York.

Cox, B.G. (1994) *Modern Liquid Phase Kinetics.* Oxford University Press, Oxford.

Pilling, M.J. and Smith, I.W.M. (editors) (1987) *Modern Gas Kinetics.* Blackwell, Oxford.

**Section G**

Atkins, P.W. (1996) *The Elements of Physical Chemistry,* 2nd edn. Oxford University Press, Oxford.

Aktins, P.W. and Freidmen, R.S. (1997) *Molecular Quantum Mechanics,* 3rd edn. Oxford University Press, Oxford.

Cox, P.A. (1996) *Introduction to Quantum Theory and Atomic Structure.* Oxford University Press, Oxford.

Woodgate, G.K. (1980) *Elementary Atomic Structure.* Oxford University Press, Oxford.

Gasser, R.P.H. and Richards, W.G. (1988) *Entropy and Energy Levels.* Oxford University Press, Oxford.

**Section H**

Murrell, J.N., Kettle, S.F.A. and Tedder, J.M. (1985) *The Chemical Bond.* Wiley, Chichester.

Coulson, C.A. (revised R. McWeeny) (1988) *The Shape and Structure of Molecules.* Oxford University Press, Oxford.

Mcweeny, R. (1979) *Coulson's Valence.* Thomson Litho Ltd, East Kilbride.

Aktins, P.W. and Freidmen, R.S. (1997) *Molecular Quantum Mechanics,* 3rd edn. Oxford University Press, Oxford.

Cox, P.A. (1991) *The Electronic Structure and Chemistry of Solids.* Oxford University Press, Oxford.

Kihara, T. (Trans. by S. Ichimaru) (1978) *Intermolecular Forces.* John Wiley and Sons, Chichester.

**Section I**

Banwell, C.N. and McCash, E.M. (1994) *Fundamentals of Molecular Spectroscopy,* 4th edn. McGraw-Hill, London.

Williams, D.H. and Fleming, I. (1995) *Spectroscopic Methods in Organic Chemistry*, 5th edn. McGraw-Hill, London.

Hollas, J.M. (1996) *Modern Spectroscopy*, 3rd edn. Wiley, London.

McHale, J.L. (1999) *Molecular Spectroscopy*. Prentice-Hall, London.

Brown, J.M. (1998) *Molecular Spectroscopy*. Oxford University Press, Oxford.

# INDEX